THE METAPHYSICS OF SCIENCE AND FREEDOM

The Metaphysics of Science and Freedom

From Descartes to Kant to Hegel

WAYNE CRISTAUDO
Department of Politics
University of Adelaide

Avebury

Aldershot · Brookfield USA · Hong Kong · Singapore · Sydney

© W. Cristaudo 1991

All rights reserved. No part of this publication may be reproduced, stored in a retrieval system, or transmitted in any form or by any means, electronic, mechanical, photocopying, recording or otherwise without the prior permission of Gower Publishing Company Limited.

Published by
Avebury
Gower Publishing Company Limited
Gower House
Croft Road
Aldershot
Hants GU11 3HR
England

Gower Publishing Company
Old Post Road
Brookfield
Vermont 05036
USA

ISBN 1 85628 132 9

Contents

Preface vii

PART I   CARTESIAN DUALISM 1

1   A metaphysics for building a new world 3

    The Cartesian project 3
    Advancing behind the mask 5
    Metaphysics as a foundation for physics 8
    The deceptive nature of perception 9
    The epistemic function of the Cartesian soul 11
    The problem of the unity of the soul and body 16
    The problem of the immortality of the soul 20
    The Cartesian God 21
    Cartesian freedom 29
    The supersession of Descartes's science and the consolidation of the mechanical world picture 33

PART II   TRANSCENDENTAL IDEALISM 45

1   Kant's foundational questions and their philosophical context 47

    Philosophical discourses of Locke, Leibniz, Kant and Hume 48
    The 'facts' to be explained 52
    Synthetic judgments which are *a priori*: The scientific foundations of metaphysics 56

| 2 | **The strategy and apparatus of Kant's critical philosophy** | 65 |

The transcendental ideality of space and time 66
The significance of the distinction between appearances (*Erscheinungen*) and noumena in Kant's strategy 74
Transcendental logic part 1: The acquisition of *a priori* concepts 77
Table of the transformation of the elements of the 'Transcendental Analytic' 87
The second 'Analogy of Experience': The law of cause and effect 90
Transcendental logic part 2: The ideas of reason 92

| 3 | **The transcendental foundations of absolute freedom and the site of faith** | 109 |

Rousseau: The need for a transcendent will 109
Kant's legitimation of moral freedom 112

PART III   ABSOLUTE IDEALISM   127

| 1 | **From transcendental to absolute idealism** | 129 |

Anti-dualism and Kant 129
The new metalevel: Fichte and the conditions of the transcendental conditions 131
Schelling and the absolute 137
Hegel's critical assimilation of Schelling 142

| 2 | **Hegel's absolute idealism and the primacy of the concept** | 152 |

| 3 | **Hegel's project: Reconciling the finite with the infinite** | 158 |

The problem with the Naturphilosophie 165

| 4 | **The substantiation of freedom: A metaphysics of the state** | 173 |

Conclusion   191

Index   195

Preface

The two great events which have shaped the terrain of modernity remain the scientific revolution and the centrality given to the concept of freedom in social, political and (more contentiously) economic life. The part that philosophy played in spreading the ideas of the new science of mechanics and specifying the nature and conditions of freedom may be disputed. Nevertheless, it is impossible to understand the radical nature of modern philosophy unless one grasps the association between its foundational concepts and strategies and the new science and the new spirit of freedom.

This is a study of that association as it is developed from Descartes to Kant to Hegel. The seminal cultural importance of that association is immediately evident when we consider the problems that Descartes, Kant and Hegel undertook to solve: the scientization of the world for the alleviation of human suffering (Descartes), the circumscription of the boundary of science and hence the protection of moral autonomy and reflective judgment from the encroachment of positive science (Kant), and the salvation of spirit and the protection of the 'ethical life' from the Enlightenment's two pronged attack of positivism/empiricism and moral voluntarism (Hegel).

Readers of Descartes, Kant and Hegel are often mystified by the central concepts, purposes, strategies and assumptions, particularly the epistemological and ontological ones of these three systems. The aim of this work is to clarify the linkages between the foundational and metaphysical transformations that are made as each thinker grapples with different aspects of the problems of science and freedom. It is not, then, an attempt to detail their respective conceptions of science and freedom (though this is touched upon), but to trace the connections between the metaphysical, ontological and epistemological problematics with science and freedom in Cartesian dualism, transcendental and absolute idealism.

Such an undertaking involves reviving/developing readings which may now appear provocative or not widely grasped. For example, this books draws attention to dissimulation in Descartes's writings and the *this worldly* purpose behind Descartes's argument for the existence of the soul and God. It focusses upon the inseparable connection between Kant's theory of mathematical judgments as synthetic *a priori* and mechanics, and it clarifies the rationale of Kant's dualism and why he maintains it throughout his architectonic. It also identifies the Fichtean and Schellingian roots of Hegel's thought and the affinities between Hegel and the German romantics, the decisive influence that the opposition of faith and knowledge plays in his philosophy, and the rationale behind Hegel's substitution of a *Logic* for metaphysics.

Finally, the interpretations offered throughout the book, particularly on Kant and Hegel, draw heavily upon the German scholarly tradition, and it is hoped that the approach adopted in this work will contribute to acquainting readers with something of the richness of that tradition.

There are a number of people I would like to thank for making this a better book than it would otherwise have been. Early in the project correspondences with Ruby Meagre made me appreciate the subtleties and strength of the Kantian conception of moral freedom. Greg O'Hare, Jann Holl, Ulrich Pätzholdt, Brent Gregston were an important stimulus to my work, and Ian Hunt and Janusz Sysak read earlier versions of the manuscript and made useful suggestions. Gyorgy Markus and Eugene Kamenka, as Ph. D. examiners, made useful critical comments on an earlier version of this work. I was fortunate to have the University of Adelaide provide me with financial assistance to spend time in the Philosophy department at the University of Freiburg. I have been fortunate to have had such a supportive work environment as the Politics department of the University of Adelaide. I would also like to thank Chris Hill and, especially, Ruth Ellickson for their help in the preparation of the manuscript, and Robert Martin and Patrick Bishop for proof-reading part of the manuscript. Two people, both excellent teachers, deserve special thanks for their friendship, guidance and support. Paul Corcoran supervised the Ph. D. out of which this book grew. His critical guidance and demand for clarity were invaluable. The book is dedicated to my wife, Cathy. Without her, this book would probably not have appeared.

The usual disclaimers about errors and disagreements apply.

PART I
CARTESIAN DUALISM

1 A metaphysics for building a new world

The Cartesian project

In *Discourse on the Method of Rightly Conducting One's Reason and Seeking the Truth in the Sciences* Descartes unambiguously expressed the fundamental purpose of his philosophy. It was to contribute to the discovery of knowledge that was 'useful in life', and to promote the 'general welfare of mankind.'[1] Descartes described the philosophy of the *Discourse* as a 'practical philosophy.'[2] The work was not to serve as a self-contained treatise, as it is now studied in universities throughout the world, but as a prefatory work to studies on optics, geometry, and meteorology. Nor was the *Discourse* to be considered, as it is now, as primarily a text in speculative philosophy. It was intended by Descartes to 'replace the speculative philosophy taught in the schools.'[3] In attempting this task Descartes made claims for his philosophy which were as spectacular as they were enticing:

> Through this philosophy we could know the power and action of fire, water, air, the stars, the heavens and all the other bodies in our environment, as distinctly as we know the various crafts of our artisans; and we could use this knowledge — as the artisans use theirs — for all the purposes for which it is appropriate, and thus make ourselves, as it were, the lords and masters of nature. This is desirable not only for the invention of innumerable

devices which would facilitate our enjoyment of the fruits of the earth and all the goods we find there, but also, and most importantly, for the maintenance of health, which is undoubtedly the chief good and the foundation of all other goods in this life. For even the mind depends so much on the temperament and disposition of the bodily organs that if it is possible to find some means of making men in general wiser and more skilful than they have been up till now, I believe we must look for it in medicine... I am sure there is no one...who would not admit that all we know in medicine is almost nothing in comparison with what remains to be known, and that we might free ourselves from innumerable diseases, both of the body and of the mind, and perhaps even from the infirmity of old age, if we had sufficient knowledge of their causes and of all the remedies that nature has provided.[4]

Although radical in its implications for social life, there was nothing particularly new about fostering interest in practical science. The growth of interest in the practical sciences had already spread far in Europe. Major technical advances had been initiated in the Middle Ages and developed in the fourteenth, fifteenth and sixteenth centuries. The mill, the spring, the pedal, printing, as well as innovations in navigational techniques, ship-building, mining, the war industry and water transport illustrate at once the extent of scientific achievement that had been made by technicians prior to Descartes.[5] In addition, there is no doubt that in astronomy and physics generally, Kepler and Galileo, and in medicine Harvey, made specific contributions which (with the exception of Descartes's discovery of analytical geometry) were more important to science than anything proposed by Descartes.[6] Even the public project outlined by Descartes had been anticipated, by Bacon. Yet Descartes's influence on seventeenth century science outweighed that of all his contemporaries. It was, as Fredrick Nussbaum pointed out, like a 'tidal wave, engulfing and sweeping along followers, critics and opponents together.'[7]

What was novel was the rare combination of the promise of a vastly improved quality of every-day life and a method not only wide ranging and rigourous enough to command the attention of the finest scientific minds of the age, but sufficiently clear and certain in its most basic premises to be grasped by almost any educated person. And it was the educated public he wished to reach with the *Discourse*. Before settling on the title *Discourse On the Method of Rightly Conducting One's Reason and Seeking the Truth in Sciences* Descartes had contemplated an unwieldy but even more enticing title for the public, *The Plan of a Universal Science to raise our nature to its Highest Degree of Perfection, with the Dioptrics, the Meteors and the Geometry;* in *which the most Curious Topics which the Author has been able to choose to give Proof of his Universal Science are Explained in such a Manner that even those who have Never Studied them can Understand them.*

There were two main reasons why Descartes wanted to capture the hearts and minds of the educated public. Firstly, he wanted to win over the artisans. Their skills were to be systematically integrated with the new physics. Secondly, Descartes knew that there were many people whose belief systems and livelihoods were dependent upon the continued dominance of scholastic Aristotelianism within the church and universities. The good will of the public was to be an important ally in the war against the Aristotelians who were doing all in their power to prevent the spread of what they saw as the pernicious anti-Aristotelian ideas being spread by thinkers like Galileo and Descartes.[8]

Advancing behind the mask

It is impossible to grasp the coherence and deeper meaning of Descartes's philosophy unless one appreciates that a major strategy in Descartes's war against the Aristotelians was the adoption of a mask of subterfuge, irony and contradiction. It is behind that mask that his philosophy advances.

That he saw it as necessary to adopt such a strategy to deal with his opponents is clear from an early fragment where he says:

> Actors, taught not to let any embarrassment show on their faces, put on a mask. I will do the same. So far, I have been a spectator in this theatre which is the world, but I am now about to mount the stage, and I come forward in a mask.[9]

The same sentiment was expressed years later to his friend and disciple Mersenne, when he wrote: 'I desire to live in peace and to continue the life I have begun under the motto *to live well you must live unseen.*'[10]

One of the most conspicuous examples of this masking can be seen in the contradictory manner in which Descartes enunciates his project and new method in the *Discourse*. Descartes presents himself in the *Discourse* as reconciled to the present order of things, while simultaneously advancing a method which requires a radical overhaul of what exists. This contradiction is obvious to any attentive reader in the following passage where Descartes eschews reform. In discussing the prospect of pulling down an old house in order to build a more comfortable abode, Descartes claims:

> Admittedly we never see people pulling down all the houses of a city for the sole purpose of rebuilding them in a different style to make the streets more attractive; but we do see many individuals having their houses pulled down in order to rebuild them, some even being forced to do so when the houses are in danger of falling down and their foundations are insecure. This example convinced me that it would be unreasonable for an individual to plan to reform a state by changing it from the foundations up and overturning it in order to set it up again; or again to plan to reform the body of the sciences or the established order of teaching them in the schools.[11]

The last sentence gives the game away. For it is precisely the reformation of the sciences that the is at the centre of the Cartesian project. The same duplicity is evident in another passage where Descartes is writing about public institutions:

> These large bodies are too difficult to raise up once overthrown, or even to hold up once they begin to totter, and their fall cannot but be a hard one. Moreover, any imperfections they may possess — and their diversity suffices to ensure that many do possess them — have doubtless been much smoothed over by custom; and custom has even prevented or imperceptibly corrected many imperfections that prudence could not so well provide against. Finally, it is almost always easier to put up with their imperfections than to change them, just as it is much better to follow the main roads that wind through mountains, which have gradually become smooth and convenient through frequent use, than to try to take a more direct route by clambering over rocks and descending to the foot of precipices. That is why I cannot by any means approve of those meddlesome and restless characters who, called neither by birth nor by fortune to the management of public affairs, are yet forever thinking up some new reform. And if I thought this book contained the slightest ground for suspecting me of such folly, I would be very reluctant to permit its publication. My plan has never gone beyond trying to reform my own thoughts and construct them upon a foundation which is all my own.[12]

Again, it is the last sentence which alerts the reader to the irony, and double speak of the author. The private nature of the project sits well with the self-effacing person who earlier suggests that his wisdom is perhaps a mere trifle, 'a bit of copper and glass'[13], which he may have mistaken for gold and diamonds. But it is in striking contradiction with the person who likens himself to the commander of armies and who holds out the promise of making humans lords of nature if they but follow his method, pool their results, communicate them to the public and build successively upon the work of past generations.[14] Likewise, we need to be on our guard, when the same man who wants to control nature in order to make us healthy also teaches in a 'provisory' maxim that we should make a virtue of necessity, so that 'we shall not desire to be healthy when ill or free when imprisoned, any more than we now desire to have bodies of a material as indestructible as diamond or wings to fly like the birds'.[15]

In spite of Descartes's public criticisms of reformers, one must realize that this criticism is to throw his enemies off the track, for the success of the Cartesian project necessitated reform. Firstly, censorship laws would have to be relaxed considerably so that scientists such as Galileo (who is alluded to at the commencement of the sixth 'Discourse') and Descartes himself could communicate *all* of their ideas to each other and to the public without any fear. Descartes did not want to experience a similar fate to Galileo and throughout the *Discourse* Descartes intimates that he is afraid of releasing some of his scientific insights to the public.[16] His close associates also knew that he had suppressed publication of *The World* out of fear of possible repercussions. Secondly, reform in the universities was essential if the new science were to advance.[17] Finally, there was no reason why the Cartesian should accept any institutional impediments to the project, if those impediments were simply the product of traditional prejudice.

In the *Discourse*, Descartes himself had set an example of the provisional nature of the loyalty of the Cartesian toward traditional mores and institutions. Shortly after having publicly sworn allegiance to the laws and religion of his country, he adds that it would be a sin against good sense to feel bound at a later date by a judgment that is no longer applicable.[18] That the maxim of allegiance is designed to ward off opponents is amply evident from Descartes's comment about the provisory maxims to Frans Burman, that he was

> compelled to include these rules because of people like the schoolmen; otherwise, they would have said that he was a man without any religion or faith and that he intended to use his method to subvert them.[19]

Descartes does not only mask the political and social requirements and consequences of his teachings. Many of his most radical ontological, epistemological and metaphysical claims are also advanced behind a mask of orthodoxy, as he attempts to introduce a new physics while simultaneously subverting the Aristotelianism which he believes is such an impediment to human progress. It is important to bear this in mind as we consider the close association between Cartesian physics, ontology, epistemology and metaphysics.

Metaphysics as a foundation for physics

In *The Principles of Philosophy*, Descartes says that metaphysics contains 'the Principles of knowledge; among which is the explanation of the principal attributes of God, of the immateriality of our souls, and of all the clear and simple notions which are in us.'[20]

Apart from the running together of epistemological and metaphysical issues, the definition of metaphysics is straight forward enough. Concern with the attributes of God and soul are also part of the traditional concern of rationalist metaphysics. But the function of metaphysics has undergone a radical transformation in Descartes's hands. The task of metaphysics is to provide sustenance to physics, which is the source 'of all other branches of knowledge', which he in turn reduces to three — medicine, mechanics, and ethics. In other words, metaphysics is essentially an instrumental, or to use Descartes's term, 'practical' science. Stated otherwise, Cartesian science makes of God and the soul, not the final things of contemplation in the ascending order of being, but the first principles to be grasped for the mastery of nature. That metaphysics is a propadeutic to physics is not always so obvious in Descartes. For example, in the 'Dedicatory letter to the Sorbonne' which opens the *Meditations*, Descartes suggests that he is engaging in metaphysics to convert disbelievers.[21] Yet, as I shall argue in more detail below, even in the *Meditations*, the metaphysics serves the cause of his physics. So much so that Descartes was to advise Frans Burman not to waste too much time on metaphysics:

> A point to note is that you should not devote so much effort to the *Meditations* and to metaphysical questions, or give them elaborate treatment in commentaries and the like. Still less should one do what some try to do, and dig more deeply into these questions than the author did: he has dealt with them quite deeply enough. It is sufficient to have grasped them once in a general way, and then to remember the conclusion. Otherwise, they draw the mind too far away from physical and observable things, and make it unfit to study them. Yet it is just these physical studies that it is most desirable for men to pursue, since they would yield abundant benefits for life. The author did follow up metaphysical questions fairly thoroughly in the *Meditations*, and established their certainty against the Sceptics and so

on; so that everyone does not have to tackle the job for himself, or need to spend time and trouble meditating on these things. It is sufficient to know the first book of the *Principles*, since this includes those parts of Metaphysics which need to be known for Physics and so on.[22]

That the author of the *Meditations* should hold such a view may seem surprising. But if one bears in mind that Descartes's project is first and foremost that of advancing physics, then it makes much sense. As does his comment to Mersenne about the *Meditations*:

> and I may tell you, between ourselves, that these six *Meditations* contain all the foundations of my *Physics*. But please do not tell people, for that might make it harder for supporters of Aristotle to approve them. I hope that readers will gradually get used to my principles, and recognize their truth, before they notice that they destroy the principles of Aristotle.[23]

Why metaphysics should serve as a foundation for physics, and why Descartes thought he could consolidate his physics while undermining the Aristotelians by advancing a metaphysic is not obvious, at least not until we have a better grasp of the problem that the metaphysics is meant to solve.[24]

The deceptive nature of perception

Descartes's ontology and hence Cartesian physics has its starting point in the simple observation that the raw data perceived by the senses is not meaningful of itself. It is always classified and the ways of classifying data vary from the primitive classifications which suffice for the simple routines of daily life to the more complex classifications resulting from scientific inquiry.

The most primitive classification is, for Descartes, based upon the pain and pleasure, harm or benefit associated with data. The degree of reality attributed to objects, says Descartes, is initially based on the power of the sensations made upon us.

> And relating all things solely to the utility of the body in which it was immersed, the mind thought that there was more or less substance in each object which affected the body, accordingly as the body was more or less affected by that object. As a result, the mind thought that there was much more substance or corporeality in rocks or metals than in water or air; because it perceived more hardness and weight in the former. Indeed it esteemed the air as absolutely nothing, as long as it experienced in it no wind

or heat or cold. And because no more light shone upon it from the stars than from the tiny flames of lamps; it accordingly represented stars to itself as being no larger than those flames. And because it did not note that the earth was rotated or that its surface was curved like a globe, it was therefore more inclined to think both that it was immobile and that its surface was flat. And our mind has been filled from earliest childhood with a thousand other prejudices of this kind; which it subsequently, in youth, did not remember having adopted without sufficient examination but accepted as most true and evident; as if known by perception or imparted to it by nature.[25]

The view of nature that is built upon what Descartes calls the prejudices of our childhood and which is described here is that which he ascribes to the scholastics.[26] According to Descartes, the scholastics take ideas such as sound, heat, moisture etc. as if they are perfectly clear. Because they feel hot, cold, moist etc. they believe that there must be substances which possess the qualities we feel — there must be a hot substance, a cold substance, a moist substance etc. The world is composed of a multiplicity of discrete substances. And the job of the student of nature is to identify and define them. There is no suggestion that the differences between physical attributes may be of degree, rather than kind. From this perspective it is impossible that the differences between physical qualities are essentially quantitative.

An important consequence of this way of thinking, especially when it is governed by the belief that all physical beings and properties are essentially purposeful, is that the study of nature has no instrumental value. Nature is not intended to be understood instrumentally — it is primarily an opportunity for enhancing speculative thought. Nor is the scholastic method with its cataloguing of substances capable of identifying the kinds of physical relationships which make instrumental controls possible.

Now even at the level of common sense, we know that a number of ideas we believe derive from our immediate sense data do not correspond to a more informed understanding of the data. To take some examples used by Descartes, we know that it is possible for someone with jaundice to believe that all objects are yellow, it is possible for someone who has lost a limb to feel pain where the limb no longer exists, and it is possible to believe that stars are tiny.[27] But we all recognize that these judgments are wrong. In that recognition we acknowledge that the adequacy of the judgment depends on the correct use of our reason. The problem that has to be solved if we want to maximize the significance of sense data is: what is the correct use of our reason? It is the search for the answer to the problem of 'rightly conducting one's reason' which connects the formation of a first philosophy or metaphysics with the development of a new physics.

And the search for that answer commences with a sweeping away of false interpretations of sense data, a break with the prejudices of scholasticism and childhood. Such a break requires subjecting all our judgments of nature to doubt, and then commencing again with firm and indubitable first principles. Only then will we have a genuine science of nature. But where are indubitable first principles to be found? Descartes's answer is that they must be found in thought itself. More precisely, they must be based on a thought which is itself indubitable.

The epistemic function of the Cartesian soul

As every philosophy fresher knows, that indubitable thought, for Descartes, is the thought of the I. For, argues Descartes, even in doubting everything it is impossible to avoid the realization that I must be a thinking being who doubts. Even if I were deceived in everything, there must be an I to be deceived.[28] For Descartes this I is nothing other than thought; doubt is merely one of its modes. While it is possible to doubt the truth of what I think, thought itself cannot be doubted. For doubting is an act of thought. Likewise, when I think of what I am, I am enmeshed within the modes of thought, i.e. sensing, imagining, understanding, doubting, etc. Thought and its modes exist because I exist, and I know that I exist because they exist. It is important to bear in mind, how stark this Cartesian I is. It is nothing other than these operations of thought.

Philosophers may well be unconvinced by Descartes's swift abandonment of doubt once he has found the *cogito*.[29] For all of the modes of thought require further elucidation, as does the concept I. Moreover, the fact that Descartes is indifferent to the substantive content of this I, to its language, its class, its gender, its history etc. makes Descartes's acceptance of the I as a clear and distinct foundational principle pretty hard to swallow. Yet, the single mindedness of the project, the instrumental purpose of the Cartesian strategy indicates that Descartes has no interest in further doubting the modes of thought.[30] He can presume that while all his readers understand very little about the mechanical structure of the world and only a few comprehend the principles, methods, and mathematics required to unlock those structures, they all know as much as they need to about affirming, willing, imagining etc. for the purpose of model building and scientific thinking.

For Descartes, two distinct kinds of activities may be distinguished when we think about thought: the activities of understanding and willing. The operations of willing include desiring, avoiding, affirming, denying and doubting. The modes or faculties of understanding are: (1) the pure intellect or pure understanding; (2) the imagination; and (3) sensation. All of these faculties prove to be essential for the Cartesian. But because sensation can be misleading, the Cartesian requires moving from the least obscure ideas, i.e. those which exist within the mind itself to the more obscure, i.e. sensation. The movement is the antithesis of the Aristotelian procedure. One does not move from sensation to the rules and ideas of existence, rather sensation must conform to rules and ideas existing within the mind of the subject. Stated otherwise, sensation is comprehensible only once the rules underpining the distinct sensations have been clarified. Note that Descartes is emphasising plan not in order to replace information that only experience can yield, but to maximize the information that can be gathered from the data.[31] To understand the mechanics of sensation one must understand what transforms obscure data into a series of relations. For Descartes this requires a dualist philosophy, and it involves the greatest triumph of Cartesian science over Aristotle, the founding of analytical geometry.

We are able to see how analytical geometry and Cartesian science generally depends, for Descartes, upon the clear distinction between mind and body when we consider the claim in the *Meditations* that we are able to classify all bodies as essentially extended because we are thinking beings. To clarify what he means Descartes uses the example of a piece of wax. The wax has colour, figure, size, it produces a sound when struck with the finger, it is hard and cold, easily handled. Similar properties can be perceived in numerous bodies. Yet when the wax is heated these properties disappear. It becomes softer, its temperature changes, its colour changes, it no longer produces a sound etc., but the wax itself does not disappear. The only thing that remains is 'the fact' that it is extended, flexible and movable. No matter how many shapes one makes of the wax, all of them require the existence of the extended substance. What, asks Descartes, perceives the essence of the wax, and is able to avoid being misled by the plethora of associations and changing properties? He answers that the understanding alone perceives that the essential nature of the wax is extension.[32] The example of the wax applies, for Descartes, to bodies in general. They are essentially extended, and it is our mind which perceives extension as the essence of bodies.

The importance of bodies being defined as extended holds the key to the Cartesian conception of the universe as a huge machine composed of mobile geometrical shapes. For if a body is defined by virtue of its extended nature, it must also be measurable. An extended thing must have a shape. Moreover, shape itself, as Descartes had demonstrated in the (posthumously published) *Rules* and in the *Geometry* can be represented algebraically and thus numerically. Instead of defining colour, hardness, heat etc. as substantial entities composed of equally mysterious qualities which we foolishly think we comprehend, we can, believed Descartes, reduce all physical existents to the strictest science of all, mathematics.[33] Mathematics is the strictest science because its truths are axiomatic, and its errors can easily be perceived by anyone who has taken the pains to master specific operations, it is applicable to bodily things and its application is useful for improving our lives. In the language of the Cartesian, the ideas of mathematics are clear and distinct.

Once we reduce all bodies to extension we have a picture of a universe as a vast continuum of moving corporeal shapes, bodily motions which, according to our different sense organs, are interpreted as colour, weight, hardness, pain etc. Note that the human body is part of this continuum, itself a part of the great machine, and to be studied, as it is in detail in the *Optics*, and *The Treatise on Man*, as a machine.

Within the framework of the universe as a machine composed of the moving shapes of one indefinitely extended substance — i.e. its parts are capable of indefinite geometrical division, depending upon which geometrical relations are to be observed — it is not fanciful to talk of the more immediate properties of colour, temperature etc. as secondary, or even illusory; number, figure, position, duration and motion are seen as primary, as the real properties. There is nothing mystical in this. Attention is merely being drawn to the fact that within the more sophisticated scientific representation of reality, the everyday sensory classification are epiphenomena governed by complex mechanisms which have to be understood if we are to subordinate nature to our will. The real mechanisms have little in common with the every-day associations we attribute to the immediately perceived properties. If we only had our senses and the faculty of the imagination we would not be able to discover these mechanisms, because we would not be able to comprehend 'the essence' of matter, nor to carry out the mathematical operations which make it possible to have real knowledge of physical qualities. This is why Descartes writes,

> I now know that even bodies are not strictly perceived by the senses or the faculty of imagination but by the intellect alone, and that this perception derives not from their being touched or seen but from their being understood.[34]

Within the context outlined above it is evident why Descartes defended his teaching from such thinkers as Hobbes, Gassendi, and his former pupil, Regius, who all saw Descartes's dualism as irrelevant or an impediment to scientific progress. For them mind was essentially explicable when it was reduced to corporeal motions. Hobbes wrote to Descartes: 'You say "I am a thinking thing"; but how do you know that you are not corporeal motion, or a body, which is in motion?'[35] And Gassendi wrote that he was unconvinced that the mind is separate from the body. Descartes, he said, had not proved that his soul is not some subtle material substance, a fire, a wind, or breath etc. For Gassendi, Descartes's own discussion of the soul being united to the body in the sixth 'Meditation' was proof that Descartes himself cannot maintain that he is an unextended thing.[36] Finally, even one of Descartes's disciplines, Regius (or Le Roy) held that the mind is not really distinct from the body, that the motions of the body determine our thoughts, that there are no innate ideas, and that 'all common notions which are engraved in the mind have their origin in observation of things or in verbal instruction.'[37]

There is a fundamental ambiguity in Descartes's teaching of the mind which explains why Regius thought that he had been spreading his master's teaching. It is worth pausing upon this for it touches the core issue that both divides and unites Descartes with the corpuscular philosophers.

The ambiguity about the nature of the mind or the soul revolves around the fact that Descartes insists on treating the soul as a substance separate from the body, thus appearing to the corpuscularists to be still entrenched in scholasticism, while persisting in discussing the union of the soul with the body. The discussion of union made it incomprehensible to the corpuscularists why there was any need to separate it in the first place. To Descartes's consternation, Regius dispensed with any need for innate ideas, i.e. ideas which are not derived from a particular sensation, nor from a combination of sensory information by the imagination. A similar interpretation was held later by La Mettrie who wrote of Descartes's dualism:

> the distinction of the two substances, thus is plainly but a trick of skill, a ruse of style, to make theologians swallow a poison, hidden in the shade of an analogy which strikes every body else and which they alone fail to notice.[38]

Regius did not realize that in doing this he had destroyed the distinction between empirical data and the epistemic ideas which enable the Cartesian to organize the data. The failure of Regius touches the major ambiguity in Descartes's teaching about the mind. There are in Descartes two types of discussions about mind which the reader must take the effort to distinguish. In the one the mind is being treated as a substance completely separate from extension, in the other the union between mind and body is being discussed. On the one hand we have an epistemological discourse, on the other a physiological one. I shall leave the latter case to one side for the moment. In the former case all of the criticisms made by Hobbes, Gassendi and Regius fail to address the epistemological significance of the dualism being advanced by Descartes. The failure is still a common-place in thinking about Descartes.

That Descartes's dualist ontology serves a crucial epistemological function is clear when we consider more closely the nature of innate ideas. There are a number of innate ideas which are crucial to Descartes first philosophy. They include, thought itself, God, and the rules of thinking correctly about nature. These latter rules, the fundamental steps of Cartesian inquiry, are as follows. Firstly, the inquirer must break all complex ideas into their most simple elements. Secondly, things are not to be treated as isolated realities, but as components of a series. In other words it is necessary to study relationships, not isolated substances. Thirdly one must constantly assess the members constituting the series being studied to make sure no inexplicable gaps exist. Where the smallest link has been overlooked the causal chain is broken, and the result is likely to be erroneous. This also enables one to retrace one's steps and discover at which point the error was made.[39]

Note that in the above prescriptions the different faculties of the mind are involved. The understanding imposes its demand for clarity and distinctness. In addition, the method requires the employment of the senses for the selection of the immediate material to be studied, and the re-working of the material by the imagination, which co-ordinates the sense impressions with the rules of the understanding and creates the models which enable the scientist to identify the laws of nature. At each step of the way an act of will is required that affirms or declines or that doubts the adequacy of the ideas presented to the mind. It also may demand more information before assent is given that knowledge has been acquired about a particular subject matter.

The Cartesian insists that it is the deployment of these operations of the mind that transform prejudices and primitive associations which are attached to sense impressions into scientific knowledge. Without these rules there could be no science, and if one simply based one's judgments on sensation, one would not arrive at these rules. Likewise, if one wanted to explain everything purely as the product of sensation, one would not arrive at the idea that nature consists of one indefinite extended substance composed of mobile geometrical configurations. Hence it is thoroughly inappropriate to talk of the contribution of these operations to scientific thought as if their essence could be discovered by a discussion of corporeal motions, as if the understanding and its rules, or the will and its operations should be analysed within the framework of extension. Accordingly Descartes writes in *The Rules for the Direction of the Mind* that

> it is impossible to form any corporeal idea which represents for us what knowledge or doubt or ignorance is, or the action of the will, which may be called 'volition', and the like; and yet we have real knowledge of all of these, knowledge so easy that in order to possess it all we need is some degree of rationality.[40]

The epistemic function of thought itself is the clue to understanding Descartes's doctrine of innate ideas. That innate ideas ultimately refer to the mind's specific contribution to knowledge is clear from Descartes's claim to Regius that all ideas *in so far as they differ in form* from corporeal motions can be said to be innate. Even the action of generalising requires an activity of understanding which separates ideas and sensations. It is simply inappropriate to investigate whether certain corporeal motions can be found in us every time a word or an idea is used. Such an investigation would contribute nothing to the furnishing of a method for the arrangement and processing of sense data. The desire to reduce the cognitive rules, methodological procedures and research maxims to corporeal motions is to be misguided about the function played by mind and that played by matter in the acquisition of knowledge. As Descartes says:

> I would like our author to tell me what the corporeal motion is that is capable of forming some common notion to the effect that 'things which are equal to a third thing are equal to each other', or any other he cares to take. For all such motions are particular, whereas the common notions are universal and bear no affinity with, or relation to, the motions.[41]

The problem of the unity of the soul and body

Had Descartes only made statements of this type, there should perhaps have been less difficulty for the corpuscular philosophers and other readers of Descartes in seeing the epistemological scope of this type of discourse. However, Descartes also wrote about the union between the soul or mind and body, while insisting, to the amazement of many readers, that there was no intrinsic contradiction in his system, and that his metaphysical dualism was indispensable to his science. Why? The probing of the epistemological function served by dualism has already clarified why he thought dualism was indispensable to Cartesian physics. An examination of two famous letters to

Princess Elizabeth clarifies why Descartes held that his teaching on the soul was consistent.

Elizabeth had inquired of Descartes how an immaterial substance, thought, could 'determine his bodily spirits to perform voluntary actions?'[42] Descartes replied that this question 'may most properly be put to me in view of my published writings.'[43] Undoubtedly an ambiguity exists in a teaching which claims that thought and body are separate substances, and that only extended things can effect extended things, while teaching that the body and thought are conjoined. Descartes acknowledged that he had not made himself clear.

> There are two facts about the human soul on which depend all the things we can know of its nature. The first is that it thinks, the second is that it is united to the body and can act and be acted upon along with it. About the second I have said hardly anything; I have tried only to make the first well understood. For my principal aim was to prove the distinction between soul and body, and to this end only the first was useful, and the second might have been harmful.[44]

At the time of this letter the public had not seen *The Treatise on Man*, which although written between 1629 and 1633 was published posthumously; nor had Descartes written *The Passions of the Soul*. While the idea of the unity of body and soul is to be found in the *Discourse* and the *Meditations*, it is in *The Treatise on Man* and *The Passions of the Soul* that Descartes specifically probes the connection between thought and body. There he analyses the activity of the 'animal spirits' (these are corpuscular movements) and the circulation of the blood in the enactment of the will, and he describes the processes which he believed occur in the brain when thought takes place. He locates the capacity to unify sensations into thoughts in the pineal gland.[45] And this same gland is said by Descartes to be the cause of the passions of the soul. It is both affected by the agitation of the animal spirits and capable of causing agitation.[46] Because the pineal gland supposedly is at the seat of the will and the seat of unifying sensations, Descartes makes the pineal gland the seat of the soul. Here Descartes's treatment of the soul is essentially corpuscular. The analysis, however faulty, is not metaphysical but physiological. But because Descartes is discussing the attributes of the soul, it may appear as if Descartes's problem is metaphysical. In *The Treatise on Man*, *The Passions of the Soul* and the *Optics* Descartes leaves it to the reader to discern the obvious fact that he is writing as a practising scientist, not as a metaphysician or epistemologist. In the letter to Elizabeth he candidly states what he had intimated in his other works.

To Elizabeth, he distinguishes between three kinds of concepts: those that apply only to body; those that apply only to soul — and these, he says, are the most elementary concepts upon which knowledge is built; and those that apply to the union of soul and body.[47] The concepts which apply to the union are the least clear. At one moment a concept referring to extension is used, at the next, one that applies to pure thought is used. In his discussion of the union between body and soul, Descartes makes an analogy between the soul in the body and heaviness in a body. The implication here is evident when we consider that heaviness in Cartesian physics is not a genuine property of body, but an unclear notion. In other words when we think of the soul being united with the body, we possess an unclear notion of the soul. A soul is no more a *substance in* a body than heaviness is a substance in a body. Thus understood the soul is an epiphenomenon.[48] But such an understanding only refers to the union of soul and body. When one is concerned with the metaphysical ideas of the soul, i.e. with rules and maxims of inquiry, then it is meaningless to talk of the union of body and soul.

This point is made even more clearly in Descartes's next letter to her when he says he 'should have shown how it is possible to conceive of the soul as material (which is what it is to conceive its union with the body), while still being able to discover that it is separable from the body.'[49] He adds that the union of the two is something sensed, but something that the pure intellect only grasps obscurely. He then says that

> people who never philosophize and use only their senses have no doubt that the soul moves the body and that the body acts on the soul....it is the ordinary course of life and conversation, and abstention from meditation and from the study of things which exercise the imagination, that teaches us how to conceive the union of the soul and the body.[50]

And he repeats that when the mind thinks of body and soul as distinct substances, it cannot think of them as united, and that when they are thought of as united they cannot be thought of distinctly. Thus he advises her as follows:

> Your Highness observes that it is easier to attribute matter and extension to the soul than to attribute to it the capacity to move and to be moved by the body without having matter. I beg her to feel free to attribute matter and extension to the soul because that is simply to conceive it as united to the body. And once she has formed a proper conception of this and experienced it in herself, it will be easy for her to consider that the matter attributed to the thought is not thought itself, and that the extension of the matter is of

different nature from the extension of the thought, because the former is determined to a definite place, from which it excludes all other bodily extension, which is not the case with the latter. And so your Highness will easily be able to return to the knowledge of the distinction of soul and body in spite of having conceived their union.[51]

The dualism, then, is held in tact, while, at the same time Cartesian science is not antithetical to corpuscularist studies. The corpuscularist, however, unlike the Cartesian, fails to distinguish the legitimate scope of corpuscular inquiry, and conflates epistemological with physiological matters.

That the dualism and the epistemology are underlabourers to physics is once again evident in the fact that after Descartes's (unsuccessful) attempt to clarify these matters for Elizabeth, he indicates that she should not spend too much time on metaphysics. For it is harmful to spend too much time upon such questions, as this impedes the intellect from engaging with the imagination and the senses.[52] In other words, one has to get on with the business of scientific practice which cannot proceed without the rules and ideas supplied by the mind. She would be better off simply applying the Cartesian method and learning more about mechanical motions than trying to correctly sort out the metaphysical subtleties which Descartes has identified for the more philosophically minded.

The mechanistic picture of the soul to be found in *The Treatise on Man* and *The Passions of the Soul* endorses the explanation offered to Elizabeth. But there is an important difference in the function of these two classics of mechanistic philosophy. In *The Passions of the Soul* Descartes focusses on human behaviour, and he reveals himself to be a pioneer of behavioural psychology. He presents the soul as physically affected by circumstances. Then he argues that recall of particular circumstances induces corporeal movements in the brain generating predictable psychological responses. To master the passions, and thus to master our responses to circumstances, Descartes proposes a number of exercises that are supposed to 'strengthen' the soul.

In *The Treatise on Man* Descartes's major concerns are anatomical. The human body is a machine, and in so far as the soul and body interact and can be analysed as an extended substance, the mind too can be studied, and is studied mechanistically by Descartes. *The Passions of the Soul* is no less mechanistic than *The Treatise on Man*, but in the former Descartes also takes care not to conflate psychological categories with corporeal motions. One must ever keep in mind the appropriate scope of the categories of inquiry. Depending upon which viewpoint one is working from the same phenomenon can be read off as bodies in motion, or as psychological states.[53]

The faculties of consciousness can also be analysed as conditions of consciousness, or bodily motions. But how one reads them demarcates the scope of the theoretical discourse one is engaged in. Descartes left it to his readers to discern that behind the idea of the soul are many diverse ideas which he employs in different types of discourses.

The problem of the immortality of the soul

Had it not been for one other idea in Descartes's writings on the soul, there may have been little difficulty in putting the pieces of Descartes's statements of the soul into their appropriate places. What caused confusion was the retention by Descartes in the *Discourse* of the idea of the immortality of the soul.[54] The idea was employed as if it resulted automatically from the complete separation of the soul from the body. It is easy to see why Descartes would have wanted to pay tribute to the immortality of the soul. For if Descartes's philosophy does guarantee the immortality of the soul, then the schoolmen and other religious critics would have to concede the compatibility of Cartesianism with official church doctrine. There is no doubt that Descartes did not provide any convincing proof for the immortality of the soul in the *Discourse* or *Meditations*. And in a letter to Mersenne about the *Meditations* he acknowledges this.

> You say that I have not said a word about the immortality of the soul. You should not be surprised. I could not prove that God could not annihilate the soul but only that it is not bound by nature to die with the body. This is all that is required as a foundation for religion, and is all that I had any intention of proving.[55]

The ambiguity is not overcome in this statement. For he has been content to conflate two different type of ideas in the service of religion. The immateriality of the soul that Descartes has demonstrated amounts to nothing more than the claim that the attributes of thought exclude attributes of extension, that the attributes of the one cannot be explained by the attributes of the other. That is to say, Descartes's separation of mind and body makes sense only within an epistemological discourse. To project the dualist ontology outside of the epistemology and to infer further that the soul persists after the death of the body is an illicit move. Also relevant is the fact that the substance that Descartes completely separates from the body consists of operations which can have no clear and distinct bodily representations. The actions which contribute to one's character and personality, on the other hand, are explicable, as is obvious throughout *The Passions of the Soul*, only in a discussion involving sensations, emotions and, hence, for Descartes, bodies. Given the doctrine of absolute separation it is also not clear how the impressions that have a bodily origin can leave their impressions on the soul, which they would have to do if the idea of the immortality of the soul were bound up with the idea of the individual personality who is to be saved or damned.

Whether one thinks that Descartes was aware of what he was doing here largely depends upon whether one sees Descartes as essentially a Catholic philosopher who really wanted to save the faith from atheistic and sceptical onslaughts[56] and was, therefore, blind to this elementary philosophical blunder, or as a philosopher who was conveniently using the doctrines of the church to advance his own teachings while providing criteria for subverting not only Aristotelianism, but all teachings which did not conform to the natural light. Certainly one can find countless cases where Descartes asserts allegiance to the Church.[57] But, in light of the aforementioned provisional nature of Descartes's allegiances, it is difficult to be completely certain about the significance and sincerity of Descartes's professions of faith. This difficulty is compounded by the fact that Descartes is, as we have already seen, frequently duplicitous. Finally, Cartesianism, at least in its long term effects, was subversive to Catholicism. As Richard Popkin, an advocate of the argument that Descartes was a Catholic apologist, has correctly pointed out:

> one of the major factors, if not *the* major one, in the development of modern irreligion, was the application of the Cartesian methodology, and the Cartesian standard of true philosophical and scientific knowledge, to the evaluation of religious knowledge.[58]

It is possible that Descartes may have been oblivious to the likely consequences of applying Cartesian standards to scripture and public institutions. But whatever we decide in our conjectures about Descartes's intentions — and on the matter of the sincerity of Descartes's declarations of faith the nature of the dispute leaves us with nothing other than conjectures — the God and soul of Descartes's metaphysics are not easily grafted onto Catholic doctrine. I have shown why this is the case with Descartes's understanding of the soul, I shall now consider why this is also the case with Descartes's concept of God, as I clarify its metaphysical utility for Cartesian physics.

The Cartesian God

In the 'Dedicatory letter to the Sorbonne' which commences the *Meditations* Descartes offers a proposal to the doctors of the

Sorbonne: in return for their public endorsement of the *Meditations*, Descartes will provide lucid (but not novel) proofs for the existence of God and the soul which will be able to convince even non-believers. If he were to receive their endorsement says Descartes:

> I do not doubt that all the errors that have ever existed on these subjects would soon be eradicated from the minds of men. In the case of all those who share your intelligence and learning, the truth itself will readily ensure that they subscribe to your opinion. As for the atheists, who are generally posers rather than people of real intelligence or learning, your authority will induce them to lay aside the spirit of contradiction; and, since they know that the arguments are regarded as demonstrations by all who are intellectually gifted, they may even go so far as to defend them, rather than appear not to understand them. And finally, everyone else will confidently go along with so many declarations of assent, and there will be no one left in the world who will dare to call into doubt either the existence of God or the real distinction between the human soul and body.[57]

The doctors were not won over, and the attacks upon Descartes from the Sorbonne continued. Given Descartes's genuine desire to win over the theologians from the Sorbonne, it is hard to see why Descartes allowed his irony to extend as far as it does in the last two sentences of the above passage. Yet the irony is hardly subtle in the remark about the atheists deferring to authority and taking up the cudgels for the theologians. Moreover, the situation of the atheists seems to be exactly the same as that of the disbelievers who, as Descartes points out in the opening page of the 'Dedicatory letter', remain unconvinced by the circularity of the argument from faith to the existence of God. It is possible that Descartes was pitching his arguments for the existence of God and the soul exclusively at the theologians, and that he was not concerned whether others who were to join the cause of Cartesianism were believers or not. This is one line of interpretation, and it is a line that is sensitive to the duplicity and irony that is sprinkled throughout the text. But, just as Descartes's argument for the separation of the soul contains a crucial component of Cartesianism, so too, as I will shortly show, does Descartes's argument for the existence of God. It is this fact which complicates the thesis for those who argue that Descartes is insincere in his beliefs about God. However, just as the soul is able to play a crucial role in Cartesianism because it is defined in a very restricted and ultimately heterodox fashion, the concept of God is presented as if it were perfectly orthodox, yet the more one appreciates how indispensable God is to

Cartesianism, the more one realizes how heterodox the Cartesian God is. This should not be surprising if we take Descartes's advice to Regius as an important hermeneutical clue for understanding Descartes's metaphysics:

> I should like it best if you never put forward any new opinions, but retained all the old ones in name, and merely brought forward new arguments. This is a course of action to which nobody could take exception, and yet those who understood your arguments would spontaneously draw from them the conclusions you had in mind.[59]

That Descartes's concept of God is, on the surface, an orthodox one is clear from his definition of God:

> By the word 'God' I understand a substance that is infinite (eternal, immutable), independent, supremely intelligent, supremely powerful, and which created both myself and everything else (if anything else there be) that exists.[60]

The Cartesian God is the totality of perfections. As such, Descartes — again in a perfectly orthodox manner — argues that existence must pertain to its nature. The nub of Descartes's original claim for the truth of the existence of God is that the very attributes of God point to its existence.

The main objection that Descartes must overcome if this argument is to be plausible is that 'God' is merely a projection of the human mind. In the 'Preface' this objection is said by Descartes to be one of the two objections which define the position of atheists; the other is the attribution of human feelings to God. This latter objection poses no problem to Descartes's conception of God, because God is defined in such a manner that human feelings don't come into it. The former objection is more philosophically significant, and it is answered by Descartes as follows:

> All these attributes are such that, the more carefully I concentrate on them, the less possible it seems that they could have originated from me alone. So from what has been said it must be concluded that God necessarily exists. It is true that I have the idea of substance in me in virtue of the fact I am a substance; but this would not account for my having the idea of an infinite substance, when I am finite, unless this idea proceeded from some substance which really was infinite.[61]

Descartes's argument rests on an ontological cleavage between a totality of perfections and a lesser or imperfect complex, human beings. Human beings are defined through their finitude, in particular through the particularity of their bodies, which are subject to the motions of other bodies and hence to illness and destruction, and through the limitations of their understanding. Because of our finitude, Descartes denies that we could be the cause of an idea that extends beyond the bounds of our finitude. This part of the argument is in turn based on the scholastic principle that 'there must be at least as much <reality> in the efficient and total cause as in the effect of that cause.'[62] That is, the finite cannot be the cause of the infinite. Or more broadly, 'what is more perfect — that is contains in itself more reality — cannot arise from what is less perfect.'[63] Descartes illustrates this point with an example that refers to an important principle of his physics.

> And this is transparently true not only in the case of effects which possess <what the philosophers call> actual or formal reality, but also in the case of ideas, where one is considering only <what they call> objective reality. A stone, for example, which previously did not exist, cannot begin to exist unless it is produced by something which contains either formally or eminently everything to be found in the stone; similarly, heat cannot be produced in an object which was not previously hot, except by something of at least the same order <degree or kind> of perfection as heat, and so on. But it is also true that the *idea* of heat, or of a stone, cannot exist in me unless it is put there by some cause which contains at least as much reality as I conceive to be in the heat or in the stone.[64]

We wildly misinterpret the crucial point about the cause containing as much reality as the effects, either formally or eminently, if we fail to keep in mind the major Cartesian innovation of the 'Second Meditation', that qualities and even natural beings are epiphenomenal.[65] What is real about heat or a stone is not to be grasped by the idea of a hot substance or a form of stones. Rather, heat and stones are simply extension, specific geometrical motions. Moreover, just as we must bear in mind that the reality to which Descartes is referring when he speaks of heat and stones etc. is nature in its quantitative aspect, whenever Descartes is talking metaphysically about physical entities we must also bear in mind that Cartesian physics is premised upon a totality in which all physical existents are bound together in a single series. There are no gaps — no vacuums — which interrupt the causal chain.[66] Our understanding may not grasp the entirety of the series, but this is no indication that the world itself lacks order. Rather it is simply a problem stemming from the finitude of our understanding.

This principle stands in the closest association with an attribute of God that is central to the *Meditations* and Cartesian physics, God's veracious nature. God, says Descartes, would be lacking in perfection were he to deceive:

> For in every case of trickery or deception some imperfection is to be found; and although the ability to deceive appears to be an indication of cleverness or power, the will to deceive is undoubtedly evidence of malice or weakness, and so cannot apply to God.[67]

This seemingly orthodox assertion has the result that there is a necessary fit between our clear and distinct ideas and reality. If we gather knowledge solely on the basis of the ideas which are clear and distinct — and, for Descartes, the clearest and most distinct ideas are, as we have said, the axiomatic ideas which the mind itself supplies, which includes the ideas of mathematics — then we will not err.

What causes error is not any intrinsic disorder in the nature of things, but the will overreaching the understanding.[68] Note that although 'God' is used to sanction the law-governed nature of reality and the adequacy of human reason for grasping that law-like nature, it is through the correct coordination of will and understanding that truth can be grasped. The theological twist is that although God provides the guarantee of certainty, God only enters into the metaphysical picture after the certainty of the ego has been established. It is the operations of the self-conscious ego which must not only bear responsibility for truth and error, but in which all the steps to secure truth and avoid error themselves must be found. God guarantees order, but the soul and its ideas guarantee God's existence. This point is nicely brought out by Gerhardt Krüger when he compares Augustine, who also used a version of the *cogito* argument to demonstrate the existence of God, with Descartes:

> The decisive divide between Descartes and Augustine doesn't consist in the fact that Augustine did not know the problem of consciousness and of foundations, nor that Descartes would only speak of a fictive God. Rather, both see human beings before God. But this seeing, for Augustine, receives its measure from God, while in Descartes it derives from the human spirit. For the one God is the original 'truth itself'; for the other the self-conscious I.[69]

With the ego as the source of the solution to human emancipation, it is crucial for Descartes, that we acknowledge our limits and proceed on the basis of an understanding which is conscious of those limits. What we cannot do is what the

Aristotelians have done, speculate about nature on the basis of an unclear idea of perfection within the species or the specific substances under examination. The employment of final causes in physics involves embracing a multiplicity of discrete data which we fit together on the basis of imperfect ideas about perfection and conjectures about God's intentions. Our intellect is thus turned away from the methodological and mathematical ideas and the order of nature that is displayed through the indefinite series of causal relations.

Although Descartes chastises the Aristotelians for their presumption in thinking that they 'are capable of investigating the (impenetrable) purposes of God',[70] Cartesianism does not entirely relinquish the concept of immanent perfection. It simply shifts the idea from the species and genera to the totality of existence. Perfection can only be found in the whole. As Descartes says, 'For what would perhaps rightly appear very imperfect if it existed on its own is quite perfect when its function as a part of the universe in considered.'[71]

One might even say that Cartesianism does require awareness about at least two of God's intentions. God does not deceive and hence His intentions about nature are nothing other than the order of nature itself if taken in its entirety. This point is drawn in the *Meditation* in a manner which is striking in its adumbration of Spinoza's idea of substance:

> For if nature is considered in its general aspect, then I understand by the term nothing other than God himself, or the ordered system of created things established by God.[72]

Because of God's veracity, once the order has been established, God Himself will not deviate from it (although there is no limitation placed upon His will). And in the *Principles of Philosophy* Descartes says that 'all things are preordained by God.'[73] The heterodox implication of preordination and a law governed universe is drawn by Descartes himself in *The World*, a work he was too prudent to publish:

> God will never perform any miracle in the new world, and...the intelligences or rational souls, which we might later suppose to be there, will not disrupt in any way the ordinary course of nature.[74]

There is one other major role that God plays in Cartesian physics. This is His creative power. The power that God provides in creating the world is constantly required in order for the world to continue to exist. Descartes uses the doctrine of concurrence to support the idea that the world of body, not particular bodies but the extended substance taken in its entirety, will always exist. As he says in the 'Synopsis' of the *Meditations*

> absolutely all substances, or things which must be created by God in order to exist, are by their nature incorruptible and cannot ever cease to exist unless they are reduced to nothingness by God's denying his concurrence to them. Secondly, we need to recognize that body, taken in the general sense, is a substance, so that it too never perishes.[75]

The doctrine of the perpetual continuity of substance is reenforced in Cartesian physics by the equation of creation and preservation.[76] It is also reenforced by a law which is a corollary of the metaphysical principle of concurrence and a necessary condition of physical laws: the law of the conservation of the quantity of motion. Metaphysically this is defended by the claim that God 'maintains in the sum total of matter, by His normal participation, the same quantity of motion and rest as He placed in it at that time [of creation].'[77]

As the source of power, veracity, hence order and perfection (if we take perfection as a totality), it is easy to see why Descartes insisted that God is not the product of human beings. For human beings are not the source of the totality of power, motion or order. When the Cartesian God is revealed in all its starkness, it is also easy to see why Descartes was satisfied that this God existed — there is a totality of power, a total quantity of motion and strict order to nature. This complex is a necessary condition for the possibility of Cartesian physics. It is equally easy to see why his religious opponents found little in Descartes's theology to advance the cause of Christendom or Catholicism.

Firstly, this God has nothing in common with the God of scripture, who, as the writer of the second set of objections to the *Meditations* was to point out, could and did utter falsehoods.[78] Moreover, his miraculous interventions are not reconcilable with strict adherence to a law-governed nature. This prompts us to wonder whether the 'evil genius' of the *Meditations* is a composite of the Aristotelian and biblical God.

Not surprisingly, Descartes did not want to draw attention to the differences between his God and the God of Abraham, Isaac and Jacob, and, where possible, he avoided discussing the biblical God. If the bible conformed to his thought, all well and good. But when faced with contradictions between scripture and science, he would plea humility before the church and profess his faith, ready to surrender his knowledge of God's laws to faith. That his physical doctrines clashed with scripture is most obvious in the Cartesian account of creation. Rather than endanger himself by publicly denouncing the authority of scripture, he designated his scientific conjectures about creation as hypotheses which, he claimed, were certainly false.[79] One finds a similar strategy of duplicity in his defense of heliocentrism where he concocts an elaborate and misleading conception of motion to conceal his subversion of the orthodox cosmology.[80]

There may well have been Catholic doctrines in which Descartes personally and deeply believed. But none of this effects the substance of Descartes's metaphysics. Descartes's philosophy has no concern with Christ, no concern with sin or grace, nor with the role of scripture in laying down laws of moral behaviour. There is also no explicable reason why the Cartesian should pray. Pascal's criticisms of rational theology touch the core of what, from a Christian perspective, is missing in Descartes's metaphysics.

> All those who have claimed to know God and prove his existence without Jesus Christ have only had futile proofs to offer. But to prove Christ we have the prophecies which are solid and palpable proofs. By being fulfilled and proved true by the event, these prophecies show that these truths are certain and thus prove that Jesus is divine....Apart from that, without Scripture, without original sin, without the necessary mediator, who was promised and came, it is impossible to prove absolutely that God exists, or to teach sound doctrine and sound morality. But through and in Christ we can prove God's existence and teach both doctrine and morality.[81]

And as another Catholic critic of Cartesian theology, Hardouin, was to point out in *Athei detecti*, the Cartesian God is not a person, but a Being, a thing or essence, rather than a spiritual personality.[82]

Gregor Sebba has argued that not only does Descartes not offer any rules for the salvation of the soul, but he teaches a radical separation between revelation and salvation through science.

> *L'homme purement homme* has never come to know the full power of reason because he never had an infallible guide to its right use; he will never see its fruits unless he resolutely separates the business of life from the business of salvation and keeps revealed knowledge out of his scientific mind.[83]

In fact, the logic of Cartesian metaphysics goes even further: the philosophy, as opposed to Descartes's personal comments, solicits one to reject revelation completely. For a God who reveals one thing through revelation and another thing through nature would be a deceiver.

This may seem strange in light of his numerous declarations of faith. But none of these declarations of Descartes's *religious* beliefs — as opposed to the metaphysical beliefs in a totality of power, motion, order, and perfection and the mind as a distinct source of metaphysical, methodological, logical and mathematical ideas — are generated from within the philosophical system. They are mere appendages to the system, something the Cartesian can accept or reject at will. But if Descartes's disciples were to follow Descartes's own advice that they were 'not to accept any opinion in my writings or elsewhere as true, unless they very clearly see that it is deduced from true Principles', then they would reject all ideas that could not be verified by the natural light, regardless of what Descartes himself may have believed.[84]

By failing to establish any spiritual or personal relationship between the soul and God, Cartesian metaphysics ultimately leaves us with a law-governed universe and a thinking being, a unity of cognitive operations, which is able to grasp the mechanisms of the physical world. While Cartesianism has required the existence of an impersonal yet omnipresent God in order to establish the conditions of certainty for its success, it has been the Cartesian ego that has initially created the clearing for the correct assessment of the standard of truth.

Cartesian freedom

If Cartesianism represents itself as devoted to protecting Christendom, the existential vision of Cartesianism is not one in which God guides the destiny of humanity through His revealed plan. It sets us on the path of the familiar and dominant one of modernity; human beings must become masters of their fate and their environment if they are to survive and thrive. Our traditions have failed us. We have to rebuild everything again on the basis of correct methodological principles.

This vision is vividly depicted in the second of the provisory moral maxims from the *Discourse on Method* where Descartes presents the situation of travellers lost in a forest. Such travellers, counsels Descartes, ought not to wander from side to side, nor remain in the same place, but once having chosen a direction they should remain firmly with this course of action, even though it were only decided upon by chance. There is no guarantee that the travellers will reach their destination. But if they sit and wait, no one will search for them, and if they wander unguided by any principle they will probably go round in circles. Like the travellers, humanity has no sure signs to aid it in its destination. If humanity continues as it has done nothing will improve. The only option left is to establish by an act of will a determination to use one's reason and then to deliberate upon the principles for reaching one's goals. People can always make a transition from one principle to another if they find that they are definitely in error. All they need do is retrace their steps and reconsider their bearings. But when human beings find themselves in a world which seems ruled by deception and chance, they have nothing to rely upon but their own resources, their reason and their will. At the same time, deception and chance will continue to rule only in so far as one does not use one's reason. Order not chance governs the world. This conclusion of Cartesian metaphysics is an encouragement to trust our reason. But because the order is to be found in the whole rather than the parts, it is not a guarantee of permanent or benign regularities. Cartesianism offers no such guarantees. To speak theologically, the benign conditions of human existence, the stability and regularity of specific geometrical motions is an act of grace. Or, to say the same thing without adopting the language of theologians, it is sheer good fortune that any goods accrue to particular parts within the system of nature. What human beings need to do, as Machiavelli had said when contemplating the political arena, is to subdue, in so far as possible, fortune to reason.

This subduing as we have indicated already, is unthinkable without the resolute act of the will. Without the resolve to break from the oppression of tradition, without the resolve to subject traditional judgments to the natural light there could be no Cartesian project. Nor could there be any kind of judgment, for Descartes, were it not for the affirmation or negation of the will in the mind's scrutiny of the data presented to it. Ultimately, it is the act of willing which creates not only the clearing for the operation of reason, but the necessary condition for its continued application and progress.[85] The will in its turn gravitates to the truths established by reason. Once something is genuinely understood, the more necessary it is for the will to give its assent.[86] Once, for example, one understands a mathematical rule, it is futile to hold that the will is free to accept or negate the truth of the rule. Any suspension of judgment would be a mere parlour game. The mind cannot help but accept ideas that it knows clearly and distinctly to be true.

Freedom thus understood consists not simply in the arbitrary exercise of the will — although it is because the will is free that it can affirm or negate, or choose to doubt — but in the judgment that corresponds to the proper order of things, in the fit between will and reason. This teaching is lucidly set out in the *Meditations* where Descartes writes:

> the will consists simply in our ability to do or not do something (that is, to affirm or deny, to pursue or avoid); or rather, it consists simply in the fact that when the intellect puts something forward for affirmation or denial or for

pursuit or avoidance, our inclinations are such that we do not feel we are determined by an external force. In order to be free, there is no need for me to be inclined both ways; on the contrary, the more I incline in one direction — either because I clearly understand that reasons of truth and goodness point the way, or because of a divinely produced disposition of my inmost thoughts — the freer is my choice. Neither divine grace nor natural knowledge ever diminishes freedom; on the contrary, they increase and strengthen it. But the indifference I feel when there is no reason pushing me in one direction rather than another is the lowest grade of freedom; it is evidence not of any perfection of freedom, but rather of a defect in knowledge or a kind of negation. For if I always saw clearly what was true and good, I should never have to deliberate about the right judgment or choice; in that case, although I should be wholly free, it would be impossible for me ever to be in a state of indifference.[87]

Cartesianism moves within a circle of the consciousness of freedom, the consciousness of one's freedom to break with the defective judgments of the past, and the conscious prospect of a freedom to be realized by the enhanced understanding and the technical application of that understanding. The freedom to perceive the world as fundamentally different in its surface form and its essence is bound up with the freedom to negate the given and transform it into something in accordance with human want.

Descartes himself did not translate this central role that freedom plays into the explicit call for a large scale political restructuring.[88] However, not only could the lesson of Cartesian doubt be applied to everything that had no justification other than authority, but in so far as Cartesianism itself was threatened by an unenlightened social order, all reflective Cartesians were tacitly invited to speculate about (and improve upon) the social conditions under which their new science could prosper. The connections between Descartes's critique of tradition and a new politics was clearly discerned by D'Alembert who said,

> Descartes dared at least to show intelligent minds how to throw off the yoke of scholasticism, of opinion, of authority — in a word, of prejudices and barbarism. And by that revolt whose fruits we are reaping today, he rendered a service to philosophy perhaps more difficult to perform than all those contributed thereafter by his illustrious successors. He can be thought of as a leader of conspirators who before anyone else, had the courage to arise against a despotic and arbitrary power and who, in preparing a resounding

revolution, laid the foundations of a more just and happier government, which he himself was not able to see established.[89]

The same sentiments could be found in other reformers such as Condorçet, who writes that Descartes 'commanded men to shake off the yoke of authority, to recognize none save that which was avowed by reason; and he was obeyed, because he won men by his boldness and led them by his enthusiasm',[90] and Saint-Simon, who says 'Descartes eliminated every idea of revelation or blind belief.'[91] The resonances of Cartesianism can also be found in Marxism as is obvious in the following passage from Engels:

> With the seizing of the means of production by society, production of commodities is done away with, and simultaneously, the mastery of the product over the producer. Anarchy in social production is replaced by systematic, definite organisation...The whole sphere of the conditions of life which environ man, and which have hitherto ruled man, now comes under the dominion and control of man, who for the first time becomes the real, conscious lord of Nature, because he has now become master of his social organisation. The laws of his own social action, hitherto standing face to face with man as laws of Nature foreign to, and dominating him, will then be used with full understanding and so mastered by him. Man's own social organisation, hitherto confronting him as a necessity imposed by Nature and history, now becomes the result of his own free action.[92]

In spite, then, of Descartes's numerous allegiances to the church of his birth, Cartesian rationalism played a major part in the erosion of a culture which took its directions from prophecy, revelation, authority and tradition. In order to allow the natural light to guide them, humans must ultimately break with the traditions which hold them back from appropriating the new world that they themselves will create. No longer is the priest the mediator between God and the soul. He is to be replaced by the scientist who studies and plots the motions of the vast machine and who brings about the creation of a world free from the afflictions of disease and the burdens of labour. Faith in the universal church and the unlimited power of the holy spirit has been replaced by faith in the unlimited potential of positive science and the mind.[93]

The supersession of Descartes's science and the consolidation of the mechanical world picture

The extent of Descartes's importance as a scientist by the thinkers of the Enlightenment is clear from Voltaire's assessment, that Descartes had 'given sight to the blind.'[94] He had seen the errors of the ancients and the failings of the senses. This was the assessment not only of one of the most influential thinkers in Paris, but a defender of Newton against the Cartesians. What Descartes had not done, and what made Voltaire himself a disciple of Newton, was discover rigourously quantifiable and experimentally verifiable physical principles. Descartes's conception of space as a plenum precluded any non-contiguous forces, thus making it impossible for Cartesians to accept Newton's mathematically formulated and empirically applicable conception of the world as a multiplicity of forces acting across empty spaces. This had, paradoxically, also led Descartes to hold an essentially unquantifiable cosmology. All motion, according to Descartes, took place in vortices or eddies, the boundaries of which were in collision with surrounding vortices. Comparing Descartes's procedure in formulating his cosmology with that of another great mechanist, Christian Huygens, E. J. Dijksterhuis puts his finger on the key reason why Descartes was to be so quickly superseded as a physicist.

> Unlike Huygens, who formulates everything at once mathematically, Descartes hardly ever conducts a mathematical argument and always remains very vague in expressing functional dependencies. In the vortex of the sun both the period of revolution and the size of the particles of celestial matter increase with their distance from the sun, but we do not learn according to what mathematical relation this increase takes place, while no attempt is even made to find for a planet the relation between periods of revolution, density, and distance from the sun.[95]

Although the specific experiments of Descartes were nearly all found wanting, it was not the Cartesian world view, but the Aristotelian cosmology that was the real victim of the scientific advances.[96] Newton had confirmed, not refuted the essence of Cartesianism — the mechanistic vision of the world. As E. A. Burrt has observed:

> The idea of mass had been incorporated into the Cartesian geometrical machine; and its substitution for the fanciful vortices only made the world-system seem all the more rigidly mechanical.[97]

Any further metaphysical thinking would have to take as its point of departure the framework that Descartes had bequeathed.

Notes

1. *Discourse on the Method*, p. 142 in Vol. 1 of *The Philosophical Writings of Descartes* in two volumes, translated by John Cottingham, Robert Stoothoff and Dugald Murdoch, (Cambridge: University Press, 1985). Unless otherwise stated all references to Descartes's writings are to this work, hereinafter abbreviated to *Writings*.
2. *Ibid.*
3. *Ibid.*
4. *Ibid.*, pp. 142-143.
5. Hermann Kellenbenz's 'Technology in the Age of the Scientific Revolution 1500-1700' in *The Fontana Economic History of Europe: The Sixteenth and Seventeenth Centuries*, ed. Carlo Cipolla, (Glasgow: William Collins and Sons, 1974), pp. 177-267 provides an excellent account of the development in technology in the period prior to and during the scientific revolution. Also important is Boris Hessen's highly influential paper 'The Social and Economic Roots of Newton's "Principia"' in *Science at the Crossroads*, (London: Kniga, 1931). Hessen goes into details about the specific problems that the technologies of the sixteenth centuries posed and which, guided by the growth of capitalism, provided an impetus for mechanics. See esp. pp. 8-16. Although developing technologies and the part played by practical technicians should not be underestimated in an account of the climate in which the new practical philosophies of the scientific revolution flourished, it is also important not to overlook the fact that many of the scientific problems of mechanics originated long before the technologies of the sixteenth and seventeenth centuries. As A. Rupert Hall points out against the position adopted by Hessen and others 'the most interesting scientific problems of the time tended to be still traditional ones — human anatomy, planetary bodies, the fall of heavenly bodies and so forth.' *The Revolution in Science: 1500-1750*, (London: Longman, 1983), p. 23.
6. Descartes was one of the first men to recognize 'the law of refraction'. But this had been discovered independently, and unknown to Descartes, by Snell. The nature of heat as a random agitation of material particles was an idea of Descartes that was to foreshadow later conceptions of heat.
7. *The Triumph of Science and Reason:1660-1685*, (New York: Harper and Row, 1953), p. 2.

8. On receiving Father Bourdin's objections to the *Meditations*, Descartes wrote to Huygens: 'It is now a prisoner in my hands, and I want to treat it as courteously as I can.... Every day I call my council of war about it....Perhaps these scholastic wars will result in my *World* being brought into the world.' *Descartes: Philosophical Letters*, tr. and ed. Anthony Kenny (Oxford: Clarendon, 1970), 31 Jan., 1642, p. 131.
9. *Writings*, Vol. 1, p. 2. The importance of the mask in Descartes's thought has been central to many interpretations of Descartes, though it is frequently overlooked. An older work which takes the mask as the starting point for a reading of Descartes is Maxime Leroy's *Descartes, Le Philosophe au Masque*, (Paris: Rieder, 1929). Also see Jacques Maritain's Thomist attack on Descartes, *The Dream of Descartes*, tr. Mabelle Andison, (Port Washington: Kennikat Press, 1944). In a number of works Hiram Caton has argued that Descartes's message is seriously misconstrued unless one takes account of Descartes's dissimulation. See *The Origin of Subjectivity: An Essay On Descartes*, (New Haven: Yale Uni. Press,1973), 'The Problem of Descartes's Sincerity', *Philosophical Forum* 2, 1971, pp. 555-569, and 'On the Interpretation of the *Meditations*' in *Man and World* 3, 1970, pp. 224-245. The argument for dissimulation is also canvassed by Louis Loeb in 'Is there Radical Dissimulation in Descartes' *Meditations*?' in *Essays on Descartes' Meditations*, ed. Amélie Rorty, (Berkeley: Uni. of California Press, 1986). Also see Walter Soffer, 'Descartes, Rationality and God', in *The Thomist*, 42, Oct. 1978, and the chapter on Descartes in William T. Bluhm, *Force or Freedom?: The Paradox in Modern Political Thought*, (New Haven: Yale Uni. Press, 1984). Mention should also be made of Wolfgang Röd's very fine book *Descartes: Die Genese des Cartesianischen Rationalismus*, 2nd. revised edition (München: C. H. Beck, 1982) which acknowledges the role of the mask in Descartes.
10. *Philosophical Letters*, letter to Mersenne, 10 May 1632, p. 26.
11. *Discourse*, p. 117.
12. *Ibid.*, p. 118.
13. *Ibid.*, p. 112.
14. *Ibid.*, pp. 143-145.
15. *Ibid.*, p. 124.
16. *Ibid.*, pp. 141-2. This fear is also expressed in many places throughout the correspondence. See, for example, the letter to Mersenne after Descartes had learnt of Galileo's views about the movement of the earth being condemned as heretical. *Philosophical Letters*, p. 98.

17. For an account of the opposition to science in the seventeenth century within European universities, see the chapter 'Science in the Universities' in Martha Ornstein's *The Role of Scientific Societies in the Seventeenth Century*, (Chicago: Uni. of Chicago Press, 1913).
18. *Discourse*, p. 122-123.
19. Descartes, *Conversation with Burman*, tr. John Cottingham, (Oxford: Clarendon, 1976), para. 80.
20. *The Principles of Philosophy*, (Hereinafter *Principles*), (Dordrecht: D. Reidel, 1983), tr. V. & R. Miller, p. XXIV.
21. *Meditations on First Philosophy*, in Vol. 2 of *Writings*, p. 1.
22. *Conversation with Burman*, para. 48.
23. *Philosophical Letters*, letter to Mersenne, 28 Jan., 1641, p. 94.
24. Some other works which have dwelt upon the epistemic function of Descartes's metaphysical ideas, and which should be consulted are Röd, *op. cit.*, Gerd Buchdahl's chapter on Descartes in *Metaphysics and the Philosophy of Science: The Classical Origins Descartes to Kant*, (Oxford: Basil Blackwell, 1969) and Daniel Garber's 'Semel in vita: The Scientific Background to Descartes' *Meditations*', in *Essays on Descartes' Meditations*, ed. Amélie Rorty.
25. *Principles*, bk. 1, 71. See also *Meditations* in *Writings*, Vol. 2, p. 52.
26. Daniel Garber points out that Descartes's representation of scholasticism is not always fair, and that Scholastics were divided over whether a form is a substance. 'Semel in vita: The Scientific Background to Descartes' *Meditations*', pp. 110-11, see also pp. 85-91.
27. *Principles*, bk. 1, 66-68, *Meditations*, p. 26, p. 57, p. 60.
28. *Meditations.*, p. 17. The same move from doubt to the *cogito* also occurs in the *Discourse*, *Writings*, Vol 1, p. 127 and *Principles*, bk. 1, 7.
29. The discussions of the *cogito* from a logician's perspective are too numerous to mention. Almost every work on Descartes from the analytical tradition explores the logical implications and assumptions of Descartes's move. One of the more widely discussed treatments in the literature is Jaakko Hintikka's '*Cogito, Ergo Sum*: Inference or Performance?' in *The Philosophical Review*, Vol. LXXI, No. 1, Jan. 1962, pp. 3-32, reprinted in *Descartes: A Collection of Critical Essays*, ed. Willis Doney, (London: Macmillan, 1970). Kant also criticizes Descartes for having derived existence from thought. For Kant this move of Descartes blurs the distinction that the 'transcendental critique' hopes to establish once and for all, the distinction between phenomena and noumena. See *Critique of Pure Reason*, B 422-423. More of this Kantian distinction later. But it should be emphasized that Kant is as little concerned with

the context of the move within the project of presenting a new physics as are our contemporary epistemologists. For a critique of the analytical tradition's orientation toward Descartes see Hiram Caton's 'Analytic History of Philosophy: The Case of Descartes', 12, Summer 1981, pp. 273-294, in *Philosophical Forum.*
30. In *The Principles of Philosophy* he says that it would be improper to enumerate the self-evident notions in the judgment 'I think, therefore I am.' *Principles,* bk 1, 10.
31. The distinction between scientific induction and Cartesian rationalism is fallacious. It was no doubt a useful weapon for the French disciples of Newton in their attack upon the Cartesians who refused to accept the (*occult*) force of gravity, which Newton had quantified and demonstrated to be operative in the universe. While Descartes may be criticized for not being strict enough in many of his experimental procedures, he nevertheless saw induction as an indispensable 'moment' of scientific inquiry. He distanced himself from 'those philosophers who take no account of experience and think that truth will spring from their brains like Minerva.' *Rules, Writings,* Vol. 1, p. 21. For Descartes, the more complex the knowledge desired, the more one must stipulate the precise conditions one wants to test in order to discover a specific mechanism in nature. *Discourse,* p. 143.
32. *Meditations,* pp. 20-21.
33. For an excellent essay on Descartes's conception of mathematics and the metaphysics of Cartesian mathematics see David R. Lachterman, '*Objectum Purae Matheseos*: Mathematical Construction and the Passage from Essence to Existence' in *Essays on Descartes' Meditations.* Lachterman correctly points out that Descartes's concept of mathematics is constructivist and governed by a commitment to problem-solving not theorem proving, and that there are inherent technical limitations in Cartesian constructivism, see esp. pp. 440-446. Lachterman also has some pertinent remarks about Descartes's strategy of dissimulation, see pp. 435-37.
34. *Meditations,* p. 22.
35. *Writings,* Vol. 2, p. 88
36. *Writings,* Vol. 2, p. 185, 235 ff.
37. *Writings,* Vol. 1, p. 295.
38. Julien Offray de la Mettrie, *Man as Machine,* (Illinois: Open Court, 1935), tr. G. Bussey. For La Mettrie the real teaching of Descartes is his own that 'man is a machine and that in the whole universe there is but a single substance differently modified.' p. 148.

39. *Discourse*, p. 120. I have also drawn upon *Rules* to clarify points which may not be obvious from the *Discourse*, but which are vital to Cartesian metaphysics and physics.
40. *Writings*, Vol. 1, pp. 44-45.
41. *Writings*, Vol. 1, pp. 304-305.
42. *Philosophical Letters*, p. 136.
43. Letter to Elizabeth, 21 May 1643, *Philosophical Letters*, p. 137.
44. *Ibid.*
45. *The Passions of the Soul*, pt. 1, para. 27 ff., Descartes retains the idea of the separation and unity of body and soul, cf. para. 30, 31 and 32. The passages referring to the soul and the pineal gland should also be compared with *The Treatise on Man*, (Mass.: Harvard Uni. Press, 1972), tr. T. S. Hall, p. 86 and the conclusion at p. 113.
46. *Passions*, pt. 1, 34.
47. *Philosophical Letters*, p. 138.
48. Also see *The Passions of the Soul*, pt 1, 30. In the next letter to Elizabeth, Descartes says that the analogy with heaviness 'was lame.' *Philosophical Letters*, p. 142. But it would be a mistake to see this as a complete retraction of the idea he is wanting to get across. As he says, he used the analogy to emphasize the absurdity in thinking of the soul as at the same time both distinct from and united to the body. However, when employing the analogy he is not using the same kind of arguments, the same kind of discourse, as he is when he wants us to think of the distinction between body and soul. Now a quality like heaviness is always an unreal quality, an epiphenomenon, the soul is not an epiphenomenon, but a real and distinct substance, if one is dealing with the understanding.
49. *Philosophical Letters*, p. 141.
50. *Ibid.*, p. 141.
51. *Ibid.*, p. 142-143.
52. *Ibid.*, p. 143.
53. See Röd, *op.cit.*, p. 146. Also see the dissolution of the problem of interaction by L. J. Beck in *The Metaphysics of Descartes: A Study of the Meditations*, (Oxford: Clarendon, 1965), pp. 269-276. Note especially the remark about substituting the cerebral cortex for the pineal gland at p. 275.
54. *Discourse*, p. 141.
55. Letter to Mersenne, 24 December, 1640, *Philosophical Letters*, p. 87.
56. This argument has been extremely popular in one version or another. A standard interpretation of Descartes along these lines is R. H. Popkin's *The History of Skepticism from Erasmus to Descartes*, (Assen: Van Gorcum, 1960). Also see E. M. Curley's *Descartes Against the Skeptics*, (Oxford: Basil

Blackwell, 1978). That textual support can be found for these positions is indisputable. However, when matters are weighed up with the stated aim and the heterodox nature of the Cartesian doctrines one must ask, what advantage could be served for Descartes by throwing in his lot with the Church against the atheists and skeptics, and what disadvantages would follow were he considered to be an opponent of the Faith?

57. Passages where Descartes confesses his faith are too numerous to cite. One typical passage runs 'No one who really has the Catholic faith can doubt or be surprised that it is most evident that what God has revealed is to be believed and that the light of grace is to be preferred to the light of nature.' Letter to Hyperaspites, August 1641, *Philosophical Letters*, p. 113.
58. 'Cartesianism and Biblical Criticism' in *Problems of Cartesianism*, ed. Thomas Lennon, John Nicholas and John Davis, (Kingston and Montreal: McGill-Queen's Uni. Press, 1982), p. 61. The major impetus to this form of criticism was to come from the one-time Cartesian, Spinoza.
59. Letter to Regius, Jan. 1642, *Philosophical Letters*, pp. 126-127. The letters provide other clues which suggest that Descartes was happy to use old theological bottles for the new metaphysical wine. For example, on learning from Colvius of Augustine's use of the *cogito* to affirm existence, Descartes wrote to thank him, adding 'I am very glad to find myself in agreement with St. Augustine, if only to hush the little minds who have tried to find fault with the principle.' Letter to Colvius, 14 Nov., 1640, *ibid.*, p. 84. And of *The World*, he said 'I shall call it *Summa Philosophiae* to make it more welcome to the scholastics, who are now persecuting it and trying to smother it before its birth.' Letter to Huygens, 31 Jan., 1642, *ibid.*, p. 131.
60. *Meditations*, p. 35.
61. *Ibid.*, p. 31.
62. *Ibid.*, p. 28
63. *Ibid.*
64. *Ibid.*
65. This is a common error. A recent work which repeats it is John Cottingham's *Descartes*, (Oxford: Basil Blackwell, 1986), pp. 50-52. Cottingham claims that evolutionary biology must be wrong from Descartes's position because it violates the idea of causality expressed in the passage cited above. But this would only be so were Descartes not talking about the entirety of extension when he is talking about causal relations.
66. *Principles*, bk. 2, 5, 11. This idea stands in the closest relation to Descartes's conception of vortices, and space as a plenum.

67. *Meditations*, p. 37.
68. *Meditations*, p. 40.
69. Gerhardt Krüger, 'Die Herkunft der philosophisches Selbst-Bewußtsein' in *Logos: Internationale Zeitschrift für Philosophie der Kultur*, Vol. XXII, 1933, p. 232, my translation.
70. *Meditations*, p. 39.
71. *Ibid.*
72. *Ibid.*, p. 56. The same equation is made in the *Principles*, bk.1, para. 28. Spinoza takes thought and extension as one and the same substance 'comprehended now through one attribute, now through the other.' *The Ethics*, Part II, Note, Prop. VII, tr. R. H. M. Elwes, (New York: Dover, 1955), *Works of Spinoza*, Vol. 2. In his *The Principles of Descartes's Philosophy*, (Illinois: Open Court, 1905), tr. Halbert Britan, we can see how Spinoza discovers himself in Descartes when he writes of God: 'He Himself is the object of His knowledge, indeed He is that knowledge.' p. 153. Spinoza was well aware that Descartes's God could not be take as corporeal (see Descartes's *Principles*, Book 1, XXIII). But according to Spinoza 'this does not mean only that all perfection of extension is wanting in him, but only that the imperfection of extension must not be attributed to him.' p. 42. Spinoza is no less careful than Descartes to distinguish between the power of existence and extension itself. See pp. 27, 121. Where Spinoza finds Descartes to be inconsistent is in the idea of freedom of the will. Spinoza holds that the will and understanding are one and the same and that Descartes falls into inconsistencies with his own determinism. See *Ethics*, part II, prop. XLIX, and part 5, Preface.
73. *Principles*, bk. 1, 40.
74. *Writings*, Vol. 1, p. 97.
75. *Meditations*, p. 10.
76. *Meditations*, p. 33.
77. *Ibid.*, bk. 2, 36.
78. *Writings*, Vol. 2, pp. 89-90.
79. *Principles*, bk. 3, 44, 45, bk. 4, 1.
80. See , bk. 2, 25, bk. 3, 28, 30, 33.
81. *Pensées*, (Harmondsworth: Penguin, 1966), tr. A. J. Krailsheimer, p. 86 (para. 189). Pascal was reported to have said 'I cannot forgive Descartes: in his whole philosophy he would like to do without God; but he could not help allowing him a flick of the fingers to set the world in motion; after that he had no more use for God', p. 355.
82. See the description of Hardouin's position in Cornelio Fabro's *God in Exile, Modern Atheism: A Study of the Internal Dynamic of Modern Atheism, from its Roots in the*

Cartesian Cogito to the Present Day, tr. Arthur Gibson, (New York: Newman Press,1968), pp. 92-107.
83. 'Descartes and Pascal: A Retrospect' in *Modern Language Notes*, Nov. 1972, Vol. 87, No. 6, p. 100.
84. *Principles*, p. XXVII.
85. See ch. 2 of Peter Schouls's excellent book *Descartes and the Enlightenment*, (Kingston and Montreal: McGill-Queen's Uni. Press, 1989) which argues for the primacy of the will over reason in Descartes. Also see Dalia Judovitz, *Subjectivity and Representation in Descartes: The Origins of Modernity*, (Cambridge: Uni. press, 1988) p. 179.
86. *Meditations*, p. 40.
87. *Ibid*.
88. Perhaps the most explicit account on politics provided by Descartes is in a letter to Elizabeth, Sept. 1646, where he gives his assessment of Machiavelli. *Philosophical Letters*, pp. 199-204.
89. J. D'Alembert, *Preliminary Discourse to the Encyclopedia of Diderot*, tr. Richard Schwab, (Indianapolis: Bobbs-Merrill, 1963), p. 80.
90. Antoine Nicolas Condorçet, *Sketch for a Historical Picture of the Progress of the Human Mind*, tr. June Barraclough (London: Weidenfeld and Nicolson, 1955), p. 122. The ninth stage of Progress is entitled 'From Descartes to the Foundation of the French Republic'.
91. Henri Comte de Saint-Simon, *Selected Writings*, ed. and tr. F. M. Markham, (Oxford: Basil Blackwell, 1952), p. 12.
92. Frederick Engels, *Socialism: Utopian and Scientific*, in Karl Marx and Frederick Engels, *Selected Works* (Moscow: Progress,1968), p. 426.
93. In light of the radical nature of Descartes's teaching and the problem that his separation of soul and body posed for the doctrine of transubstantiation, it is not surprising that Descartes was so vigorously attacked by churchmen. Nor should it be surprising that his works were placed on the Index, that he should die in exile and his works be banned from French universities throughout the seventeenth century. (In the University of Paris, it was not until 1720, that Descartes's ideas were officially recognized.) Yet, while one section of Christendom was denouncing Descartes as an atheist, many came to his defense. See Queen Christina's defense of Descartes in Trevor McClaughlin, 'Censorship and Defenders of the Cartesian Faith in mid-seventeenth Century France' in *Journal of the History of Ideas*, Oct-Dec. 1979, p. 576.
94. *Lettrês Philosophiques*, intro. Gustave Lanson, (Paris: Libraire E. Droz, 1937), 'Sûr Descartes et Newton', Vol. 2, p. 7.

95. E. J. Dijksterhuis, *The Mechanization of the World Picture*, (Oxford: Clarendon Press, 1961), tr. C. Dikshoorn, p. 414.
96. Aristotle could find a point of re-entry into the new world picture only by huge concessions to a new set of ontological conditions, the most important being that any retention of final causes could not be at the expense of quantifiable efficient causes. Leibniz quickly saw that Aristotle could not be as swiftly dispensed with as Descartes had hoped. For Leibniz, Aristotle was useful (a) because he believed final causes were indispensable for the study of processes and (b) he opposed the idea that extension was the substance of bodies (for Leibniz, extension is a plurality of force points and force is substantial). In addition, he sought to reconcile the Aristotelian importance of the primacy of quality over quantity, a move which binds Leibniz's theory of calculus and his conception of monads, not only with each other but with Aristotle's conception of substance. See Letter to Jacob Thomasius, April 20/30 1669, *Philosophical Letters and Papers*, intro., ed. and tr. Leroy Loemker, (Dordrecht: D. Reidel, 1969 [1954]), pp. 93-100.
97. *The Metaphysical Foundations of Modern Science*, (London: Routledge and Kegan Paul, 1932 [1924]), p. 243. There were also theological dimensions to the conflict between the Newtonians and the Cartesians, as Koyré, in 'Newton and Descartes' in *Newtonian Studies*, (London: Chapman and Hall, 1965) and Hessen, *op.cit.*, pp. 34-40, have both pointed out. Ultimately Newton's conception of forces acting at a distance was able to be interpreted as leaving room for active principles besides solely material ones. Although the tenets of Cartesian cosmology were to be cast aside by Newton, many of Descartes's ideas were to foreshadow ideas of field theory. See Table 2 comparing world views in William Berkson's *Fields of Force: the Development of a World View from Faraday to Einstein*, (London: Routledge and Kegan Paul, 1974), p. 254.
In *Diderot and Descartes: A Study of Scientific Naturalism in the Enlightenment*, (Princeton: Uni Press, 1953), Aram Vartanian specifies how Descartes's influence extended to the ostensibly anti-metaphysical materialism of thinkers like D'Holbach. Vartanian points out that even the scientific minds of the seventeenth century who ostensibly owed more to Descartes's great Epicurean opponent, Gassendi, had inherited a framework which owed more to Descartes than Gassendi: 'Through the medium of the Cartesian reform, in effect, the Epicurean natural philosophy may be said to have lost its elements of mythology and its subordination to a metaphysics built largely around the evasive concept of Chance; and instead, to have been integrated into the methodical, self-corrective search,

typical of the modern era, for the fixed and determinable laws governing physical events. Descartes had made materialism, so to speak, 'scientific.' Cartesian science, based on a corpuscular and unfinalistic view of matter in motion, thus was able to convert to its standpoint the correlative parts of Gassendi's "physics".' pp. 59-60.

PART II
TRANSCENDENTAL IDEALISM

1 Kant's foundational questions and their philosophical context

In a remark to Princess Elizabeth, Descartes made a claim that clarifies the difference between Descartes's and Kant's major philosophical concerns. Reflecting upon how he allocates his time Descartes wrote:

> I can say with truth that the chief rule I have always observed in my studies, which I think has been the most useful to me in acquiring what knowledge I have, has been never to spend more than a few hours a day in the thoughts which occupy the imagination and a few hours a year on those which occupy the pure intellect.[1]

The most important activities upon which Descartes exercised his imagination were the construction of scientific models and the conducting of experiments. The amount of time devoted to thinking about the metaphysical foundations consumed only 'a few hours a year'. Perhaps this is exaggerated, but the message is plain enough where Descartes's priorities lay. With Kant matters were otherwise. His greatest intellectual labours were devoted to problems of metaphysics, and to laying foundations for metaphysics.

Unlike Descartes, Kant was neither writing for an audience whose heads had to be freed from the doctrine of substantial forms, nor for the educated members of the public in order to involve them in undertaking experiments. Nor did Kant need to enthuse about the great possibilities of mechanistic science. He was not a pioneer of the scientific revolution. He was the heir to Newtonian science writing for an academic audience. This changed environment pervades the content of Kant's metaphysics, as well as his task of establishing metaphysics as a science.[2] In addition Kant was the heir to a number of philosophical discourses, unknown to Descartes, which had arisen in conjunction with the newly established mechanistic world model. Of these discourses there are three which are of fundamental importance for Kant's project: the discourses of Locke, Leibniz and Hume.

Philosophical discourses of Locke, Leibniz and Hume

Locke

In *An Essay Concerning Human Understanding* Locke had undertaken 'to inquire into the origin, certainty, and extent of *human knowledge*, together with the grounds and degrees of *belief, opinion*, and *assent.*'[3] The first chapter of that work opens with the defiantly anti-Cartesian pronouncement 'No Innate Speculative Principles.'[4] This very title is indicative of Locke's answer to the question how 'our understandings come to attain those notions of things we have.'[5] They are either derived from external sensible operations or reflection, i.e. the mind's own internal operations upon the data originally supplied by sensible objects. In giving this answer Locke step by step proceeds to a position in which known substances are ultimately inseparable from words, and words are only signs for ideas in the mind.[6] Our knowledge is a copy of things, yet because things are particulars and we must use general terms when we communicate and when we think, our knowledge is '*the perception of the connection of and agreement, or disagreement and repugnancy of our ideas.*'[7]

The problem with Locke's position may not be immediately apparent if we restrict our knowledge to things of which we can form a picture, i.e. if we restrict our knowledge to empirical perception. However, Locke fails to explain how the apodeictic science of mathematics is not only developed axiomatically but its results projected back onto the world of facts. He fails to explain the *fit* between the axiomatic system and the sensible world which mechanics can quantify. He makes the science of number a purely logical affair once its initial concepts have been derived from experience, yet in the expansion of a logical system empirical reality is left far beyond. Rather than explaining how mechanics is possible, it would seem that such a science is impossible.

Unintentionally Locke is as much a dualist as Descartes. But his dualism has this odd feature which stands in striking opposition to Descartes: it commences with the sensory world as the immediate and certain, and concludes by seeing that world as uncertain and ultimately unknowable. On the one hand exists a realm of words and nominal ideas, on the other a world of real essences. And, as Ernst Cassirer says, 'No bridge leads from one region to the other.'[8]

Leibniz

It was Leibniz who most thoroughly examined and criticized Locke's philosophy. Leibniz's general assessment of Locke is perhaps most succinctly summed up in a letter to Nicolas Remond: 'Mr. Locke had subtlety and skill and a kind of superficial metaphysics for which he was able to secure acclaim, but he was ignorant of the mathematical method.'[9] His critique was most fully developed in *New Essays on Human Understanding*.

Like Locke, Leibniz opens his *New Essays* by discussing innate ideas. Against Locke, Leibniz affirms the existence of innate ideas. Instead of likening the mind to a *tabula rasa* he likens it to a veined block of marble. He describes innate ideas as 'inclinations, dispositions, tendencies, or natural potentialities of the mind.'[10] It is, however, the capacity of mind to deduce rational principles, particularly of a logical, mathematical and metaphysical type, that enables, for Leibniz, the expansion of knowledge about the world. Without them we would have a blur, a confused and ultimately abstract assortment of sensations.[11] To capture the infinite *differentiae*, the infinite fineness of relations within relations, worlds within worlds, requires following the intricate series which are the perceptions of a monad. The monad is, for Leibniz, a real essence, and all our ideas refer to a real essence. Hence for Leibniz, ideas do not depend upon names, as they ultimately do for Locke, but upon the infinitely connected series of 'concrete' relations. For Leibniz all ideas are interconnected. At any given moment of reflection one enters into infinity, an infinite number of possible worlds.

> there is no term which is so absolute or so detached that it does not involve relations and is not such that a complete analysis of it would lead to other things and indeed to all other things.[12]

In other words each term derives its sense from and contributes to the sense of the entire range of other terms. The more a term clarifies the inner mechanisms of things, the more substantial, the more clear it is. The clarity of any given idea depends on the clarity generated by the entire system of ideas within which it forms a momentary point of reference.

Within this infinite chain of predications, the logically possible terms are just as determinant as the empirical world. The total reality is the sum of possible worlds. The actual world is only one of the possible worlds. The essential world is the determinant of the existing world, the latter is simply the best of the possible worlds.

In responding to Locke's empiricism Leibniz ultimately 'logicises' existence. Not only do metaphysical and logical truths become indistinguishable, but so do the logically possible worlds and the actual world, which, to repeat, is just one of the possible worlds. But, asserts Leibniz, the actual world is the best possible one. Nature always works with a sufficient reason.

The epistemological difference with Locke is thus also ontological and theological.[13] For the metaphysical principle that nature always works with a sufficient reason is supported by Leibniz on theological grounds. The world is the product of pre-established harmony, the product of God's plan, and each perception point is at once immortal and the bearer of God's plan.

Kant shared Leibniz's view that Locke had failed to address adequately how axiomatic systems can be applied to experience, and how one makes the transition from truths of experience to metaphysics. Yet Kant did not follow Leibniz's flight into the more speculative consequences of the *Monadology* and *The New Essays*.[14] To the extent that Locke had tried to explain the origin of ideas, and to restrict them to the realm of the demonstrable, Kant saw in Locke a kindred spirit who would put an end to fantastical speculation. But Locke had failed in his task. Leibniz was the proof of that failure. For after demonstrating the indispensability of rational knowledge, and the inadequacy of the 'copy theory' for explaining rational knowledge, Leibniz had gone on to make metaphysical speculations that were untestable. Locke had achieved the opposite of his intention. In Kant's words, Locke 'opened a wide door to enthusiasm — for if reason once be allowed such rights, it will no longer allow itself to be kept within bounds by vaguely defined recommendations of moderation.'[15] Had Locke been more rigourous in his reflection on the human understanding and claimed less on behalf of experience, he may have taken the route opted for by Kant, and he would have blocked the opening to the land of fantasy taken by Leibniz. That route by Kant's own account owed more than anything to Hume's analysis of causation.[16]

Kant and Hume

Hume had followed Locke in his conviction that all knowledge derives originally from experience, and that we also have ideas of relations. However, Hume focussed upon the most fundamental relation of science, causality, and he sought to define its origin. He argued that our knowledge of cause and effect is not based on necessities. It is based on nothing more solid than customary associations. Scientific judgments were little more than probabilities.[17] As useful as our science may be for the moment, it has no secure foundations.

It is important to realize that Kant did *not* disagree with Hume over the issue that because a sequence of events in the world happens in the past it must happen in the future. To appreciate what divided them we have to consider the theory of science which permeates *Critique of Pure Reason*, and which is in turn the key to understanding Kant's epistemology.

For Kant, the scientific revolution was made possible by the isolation of variables and the monitoring of their interaction. Scientists, although seeking to be instructed by nature, were no longer groping in a mass of seemingly unrelated events. They were formulating questions and principles which they could test in experiments. The new scientist was not a pupil but a judge 'who compels the witnesses to answer questions which he has himself formulated.'[18] In other words while the scientists were inquiring into nature's laws, it was the activity of the scientists rather than anything on nature's part which held the key to the idea of the legality of nature. For hitherto nature had hid her secrets, but now she was being forced to reveal them. The question of whether nature in itself and in its entirety was ordered was, as it still is, irrelevant to most working scientists. That is a metaphysical question. It has nothing to do with monitoring specific physical interactions, or formulating and applying rules which help us predict and manipulate those interactions. Neither a scientist nor anyone else can know in any instance outside of the laboratory that the relations which form a physical law will be exactly replicated. This is because our knowledge is only partial, and the laws refer to a fixed range of variables. There may always be variables in nature which we have not accounted for in the laboratory, which when discovered may lead us back into the laboratory to reconsider the hypothetical rule, or law. Expressed in Kant's terms, we can never know the world in its entirety; it is a mere idea.[19]

Hume's scepticism sits well with the limitation of human knowledge and the indefinite range of variables and their permutations and combinations to be found in nature. Hume makes us aware that we must always be prepared to reconsider our assumptions and ideas about nature, and that we cannot assume that nature will always conform to our predictions. Thus far Kant does not differ with Hume. But Kant saw Hume's scepticism as misplaced in so far as it failed to stop at the content of laws of nature and instead cast doubt upon the notion of necessary connections, i.e. laws, themselves.[20] For having abstracted the concept of causality from the various sense data and the reflections on sense data which form a causal relationship, Hume inferred that the concept causality is as precarious as the specific relationships which, on the basis of experience, we hold to be necessary connections. Kant realized that in the case of causal relationships, it is one thing to say that the perception of a series a,b,c is merely customary; it is quite another to say that there is a phenomenon which does not stand in a causal relationship with some other phenomena. Kant observed that when something unexpected happens this is no reflection whatever on the foundations of the functional necessity of the concept of causality. On the contrary, it is precisely because this concept for understanding phenomena is not thrown off but steadfastly and, Kant holds, necessarily adhered to that we believe that there must be another variable to be accounted for. There is no guarantee of finding it, but that is another issue.

Note that Kant is not disagreeing with Hume about our incapacity to make *a priori* claims about specific causal relationships. Only experience can tell us what variables stand in a causal relationship.[21] But his starting point is that the functional and formal conditions of scientific knowledge are not to be conflated with the sense data, as Hume and, according to Kant, Locke and Leibniz also did.[22] The sense data supply the content of the law, but, asks Kant, are there necessary forms or conditions of any judgment of experience (*Erfahrung*) which purport to have objective validity? How is their necessity to be established and what are they? To answer these questions Kant adopts a strategy grounded in other closely related questions, which we must now consider.

The 'facts' to be explained

Kant is not asking if objective judgments of experience are possible. He is asking *how* such judgments are possible. He assumes the existence of natural science, as a changing, progressively expanding body of laws and concepts. Moreover he realized that these laws are (a) subject to empirical verification, and (b) their formulation and advancement is inseparable from the development of mathematics. In addition he realized that classical physics works with a number of rules which are preconditions for the possibility of data conforming to law. In Kant's terminology, they are *a priori*.

Perhaps the most fundamental *a priori* relationships are those of arithmetic and geometry. Within the context of classical mechanics Kant realizes that the concepts of homogeneity, infinity, continuity, immutability, causal inertness, i.e. the properties of Euclidean space, are not found *in* experience. Yet Euclidean geometry, (and concomitantly the concept of Euclidean space) is fundamental to classical mechanics. The axiomatic system has application, and is developed through that application. Likewise he observed that neither the one dimensionality of time nor the numerical series which we use to measure the duration of the motions of phenomena are objects *in* experience. Yet those durations, and thus the motions themselves, i.e. their rate of change or the total change in the quantities of the masses under observation, can be measured with an infinitesimal degree of accuracy.

It is in this context that the first of what Kant calls 'the main transcendental questions' needs to be understood: 'how is mathematics possible?' It stands in the closest association with the second of the main questions: 'how is pure natural science possible?'[23] For Kant, pure natural science is possible (a) because phenomena can be represented as extensive magnitudes, i.e. as geometrical shapes upon a co-ordinate system, and (b) as intensive magnitudes, i.e. as sensations possessing degrees of reality. Space and time, for Kant, are thought of as continuous, and any mass is capable of being quantified to an infinitely small degree before it reaches a vanishing point. The classical properties of space and time are, for Kant, inseparable from the measurement of the interaction of forces. The lawfulness of nature thus for Kant must take account of the space-time continuum within which all interactions of forces take place, as well as the mathematical operations which measure these forces.[24]

Whereas mathematical relations can be developed or, as we shall see later, constructed *a priori* and applied to sense data, there are other (dynamic) principles which he argues are *a priori* and indispensable for any possible objectively verifiable experience. Yet they have no function apart from the observation of experience.[25] These are classified by Kant under the title of the 'Analogies of Experience.' They are: (1) the principle of the permanence of substance, which is, for Kant, equivalent to the principle that the quantum of phenomena can neither increase or decrease; (2) the aforementioned principle of the law of cause and effect; and (3) the principle of the reciprocity of all substances in space. As the formulation 'Analogies of Experience' suggests these rules are meaningful only in relationship to things of experience, or possible experience.

These principles are not of a mathematical nature, but they are readily recognisable as essential to the reference-frame of classical mechanics. Moreover, the utility of these principles in classical mechanics was not a problem for Kant or his contemporaries, including Hume. Had Kant merely assumed these principles as necessary for the formulation of physical laws he would not have differed from any other Newtonian physicist. But when Kant asks 'how is mathematics and how is pure theoretical science possible?', he is not doing physics. By his own account he is doing ontology, although it is so closely related to what we would now classify as epistemology that confusion often arises when these terms are placed in opposition for analysing Kant's philosophy.[26] In addition, in seeking to answer these questions he sees that they are traditionally enmeshed within another type of discourse, metaphysics, which deals with the supersenuous ideas of God, freedom, and immortality.[27]

Descartes, Newton, Locke, Leibniz, Berkeley, and Hume had all in one way or another entered into a discourse about these concepts. Their doing so was not something merely appended to their questions about how we have knowledge or what knowledge we have. Rather they move from a discourse about facts to one about metaphysics, as if knowledge in the one field equipped one to answer metaphysical disputes. For example, Leibniz's monad was at once the key for understanding how we know things, as it was the key for his argument about the immortality of the soul. Similarly, Newton infers from the regular motions he has discerned in the heavens that this order is the creation of an intelligent and omnipotent Being. Newton then passes from a discourse on physics to a speculative account about the nature of this non physical 'Universal Ruler.'[28] With very different intentions and results, Hume moves from a discussion of how we know things to a conclusion which ostensibly 'consigns to the flames' whatever does not contain reasoning about number and quantity or reasoning about facts.[29] In sum these thinkers were locked into a discourse of speculative metaphysics.

Instead of rushing to take sides in a dispute about these supersensible concepts, Kant starts from the fact that there is a dispute. He does not ask whether metaphysical disputes exist. Rather he asks 'how is metaphysics possible in general?' Only after he can answer this question does he undertake to solve the disputes by answering the question, 'how is metaphysics possible as a science?'[30] The disputants are to be brought before a tribunal of scientific metaphysics.

The enmeshment of metaphysical issues with issues of natural science supplied a clue to Kant about how to solve the problem of the possibility of metaphysics. If there are *a priori* concepts and principles which are indispensable to natural science (and the aforementioned principles of natural science as well as the science of mathematics indicated to Kant that there were) then it would be quite natural for thinkers of the calibre of Newton or

Leibniz to have no qualms in accepting as legitimate other *a priori* principles which may also serve a function in scientific inquiry, and which they then employ speculatively. It would be understandable why a discourse of natural science crossed over into a discourse of rational metaphysics.

Just as the *a priori* principles of natural science were ready to hand, for Kant, the main principles of speculative metaphysics were also ready to hand. Inquiries into the nature of the soul, of the world, and of God constituted the three traditional areas of speculative metaphysics: rational psychology, rational cosmology, and rational theology. The respective ideas which Kant discerned as fundamental for these three disciplines were: (1) the existence of a simple thinking substance, which is a unity yet which relates to possible objects in space; (2) the disputes over whether the world had a beginning in time and was enclosed by space; whether there are simple substances in the world; whether there is causality through freedom; and whether there is either in or apart from the world a necessary being which can be considered its first cause; and (3) the ontological, cosmological and the physico-theological arguments for the existence of God.

In addition to the question how the study of nature could lead one to speculate upon the nature of the soul, the world in its entirety and the existence of God, Kant observed that these ideas also figured in another type of discourse, moral discourse. Moreover, he thought that one of these ideas, freedom, is indispensable for the possibility of a moral judgment. Whether there is such a thing as freedom is a debatable question. Whatever side one adopts, however, does not change the fact that the dispute is metaphysical. Freedom, understood as the capacity to act according to a rational principle, refers to a supersensible idea.

If supersensible ideas stand in a necessary relationship to judgments of natural science and judgments of freedom — and for Kant such questions as 'Does the cosmos have a beginning in space and time?' 'What is the end of a free yet sensuous being?' indicate that they do — then it will be necessary to demonstrate why this necessity exists. This is one task of a scientific metaphysics. In addition, if judgments of freedom are governed by different *a priori* rules, if they necessarily differ in form from judgments of fact, then there will be, at the very least, two different types of non-empirical principles, those that have a legitimate and necessary function in natural science, and those that deal with questions of freedom, i.e. moral principles. Are there any other *types* of judgments which employ rules that are not directly based upon and testable by empirical observation?

At the time of writing *Critique of Pure Reason* Kant thought not. Later he came to hold that judgments of taste also relied upon *a priori* principles. But he also saw an important difference between the underlying rules of aesthetic judgements, and moral and scientific judgments. In the case of moral judgments the constitutive rules of the judgment, argued Kant, can be encapsulated in an overarching law, the moral law or categorical imperative. On the basis of that law it is possible to deduce axiomatically a number of moral principles. These principles do not provide the circumstances of moral action but they can provide a guide for judgment in particular circumstances when a moral decision needs to be made. Similarly in the case of natural science it is, argues Kant, possible to deduce metaphysical principles of nature which are based upon an axiomatic treatment of space, time and motion (only the last of which, for Kant, is not a condition of a judgment of experience, but an empirical part of the content of the judgment).[31] In the case of aesthetic experience no such inferences are possible. For, Kant holds, apart from compliance with the 'rules' of taste, artistic production requires a free play of the cognitive powers and the freshness and vitality of genius if it is to be either beautiful or sublime. This is beyond the bounds of metaphysics.[32]

In sum Kant distinguished between three types of *a priori* judgments: theoretical, i.e. those which relate to phenomena and which are necessarily employed in natural scientific inquiries; practical, i.e. those which are prescriptive and are necessary for moral judgments; and aesthetic, those which relate to phenomena but which are judgments of feeling not of knowledge. He also claimed a special *a priori* status for teleological judgments. But they serve an intermediary function between judgments of freedom and judgments of necessity.

Synthetic judgments which are a priori: The scientific foundations of metaphysics

It is not merely a matter of declaring some judgments to be *a priori*. Few ever denied that there are *a priori* judgments. Logic and mathematics are undoubtedly *a priori*, even though there may be dispute about whether the original *source* of their postulates is also *a priori*. If the question merely concerned the psychological or historical origin of knowledge there would be no sound reason to disagree with the empiricists. As Kant states in the opening line of the Introduction to *Critique of Pure Reason*: 'There can be no doubt that all our knowledge begins with experience.' And the first paragraph concludes equally emphatically: 'In the order of time, therefore, we have no knowledge antecedent to experience, and with experience all our knowledge begins.'[33] These statements apply equally to a psychological account of how we come to know things as to a historical account. Mathematics, for example, was used by the Egyptians long before Euclid. But Euclid's axiomatization of geometry transformed it into a science.³⁴ As the example of Euclid indicates, Kant is concerned with the indispensability of *a priori* procedures, the necessity of *a priori* elements and principles in the formation of judgments.

But it is not just the concern with the necessity of specific *a priori* principles that is at the centre of Kant's attempt to found a scientific metaphysics. Rather it is the introduction of the idea of an *a priori* judgment which is synthetic that is the revolutionary element in Kant's undertaking. It is this concept which leads Kant to claim that he has found the key for making metaphysics a science, and to declare:

> In this enquiry I have made completeness my chief aim, and I venture to assert that there is not a single metaphysical problem which has not been solved, or for the solution of which the key has not been supplied.³⁵

A judgment is synthetic when the subject of the judgment does not 'contain' the predicate. If the predicate of the judgment is (intensionally) identical with the subject then the judgment is not synthetic but analytic. Even if on immediate inspection the identity is obscure but it can later be demonstrated that the predicate is merely explicative then the judgment is analytic.³⁶ The criterion for evaluating the validity of analytical judgments is a purely logical law, the law of contradiction. Synthetic judgments on the other hand require something extra. As the very name suggests they require an act of synthesis, an act which is responsible not merely for comprehending what is already in the subject, but for bringing together representations (*Vorstellungen*) which are not logically identical to the subject. All judgments of experience are synthetic. But this does not mean that all judgments which are about an empirical subject are always synthetic. For example 'the cat is an animal' is analytic. Or, to take two of Kant's examples 'the body is extended' and 'gold is a yellow metal' are analytic.³⁷ They are analytic because whenever the particular concept is employed there is no necessity of going beyond the subject to affirm what is in the predicate. A mere analysis of the subject will yield a true predicate. The denial of the predicate 'a body is a non-extended substance' would be illogical if and when body was by definition taken to mean an extended thing.³⁸ On the other hand to find out that bodies had the power of attraction or weight did not simply require an analysis of the concept body — indeed Descartes excluded the concept of weight from a body. It was only by empirical observation that another attribute, attractive force, could be added to the subject body.³⁹

Now if we consider the judgment 'in all changes of the corporeal world the quantity of matter must remain constant' this is clearly not an analytic judgment.[40] For the concept of all matter does not logically entail the predicate of persistency (*Beharrlichkeit*), or the principle of the conservation of matter. (By matter Kant here means, as is evident from the 'First Analogy', simply what has existence in space and time. The conservation principle applies to phenomena taken as a whole not to particular substances.[41]) Yet, Kant claims that because this judgment is not the consequence of empirical observation but the necessary condition for empirical observation, this judgment is *a priori*. The principle thus is, according to Kant, a synthetic principle which is *a priori*. Here we have not shown why Kant held this was a necessary condition of empirical observation; we have only pointed out that he held this.[42]

The task of *Critique of Pure Reason* was to stipulate exactly which judgments were synthetic *a priori* and to explain 'how synthetic judgments are *a priori* possible.' This last problem came to be formulated in the second edition as the overarching question of *Critique of Pure Reason*.[43] Kant believed that only by answering this question could the foundations of scientific metaphysics be secured. For analytical judgments always rely upon some initial information or concept. They enable explication to occur but *not* the initial acquisition of any knowledge. Hence Kant says that analytical judgments only expand our knowledge formally not materially.[44] In themselves analytic judgments cannot tell us what is real. By definition they cannot affirm contradictory concepts, but there is no reason why they cannot be about non-existent things. Only if something can be observed can we know that it is real, provided we have an adequate classifier or concept to comprehend what it is that we are observing.[45] Analytic judgments cannot provide the foundations of any science, because they do not supply a 'critical' principle which would enable us to legitimate whether the judgment is objectively verifiable, and not simply not contradictory. On the basis of initial synthetic judgments, analytic judgments can explicate the terms of the judgment. But synthetic judgments alone, for Kant, can supply legitimate foundations for any type of knowledge.

If, then, there are judgments which are synthetic and *a priori*, these would provide the foundations required for a scientific metaphysic. They would be the initial and necessary principles upon which any other metaphysical principles were based.

In sum Kant's philosophy had undertaken to stipulate that certain elements, rules and principles were synthetic, *a priori* and necessary, and he had to specify which ones they were. To do this required not simply a metaphysical inquiry, but what he calls a critical inquiry, a *Critique of Pure Reason*. This was to be a propadeutic to metaphysics. In requiring such a propadeutic Kant had taken a meta-theoretical step beyond that of any of his precursors. Unlike Descartes, Leibniz and Spinoza, Kant recognized that logic, hence the clarity and distinctness of ideas, was not a sufficient condition for the validation of metaphysics. Nor was the empiricism of Locke and Hume sufficient. Only a transcendental critique could supply the metaphysical foundations. And that meant moving beyond rationalism and empiricism.

Notes

1. Letter to Elizabeth, 28 June 1643, *Philosophical Letters*, pp. 141-142.
2. Hitherto, Kant held, metaphysics was more akin to astrology or alchemy than any science. *Prolegomena to Any Future Metaphysics that will be able to present itself as a Science*, tr. P. G. Lucas, (Manchester: University Press, 1953), p. 135. All references to Kant's *Prolegomena* are to this edition.
3. *An Essay Concerning Human Understanding*, (New York: Dover, 1959), Vol. 1, p. 26.
4. Locke's argument conflates the distinction between the logical and temporal priority of innate ideas. Descartes at times also makes this conflation. But the Cartesian method relies only upon the logical priority of innate ideas. Cassirer in *The Platonic Renaissance in England*, tr. James Pettegrove, (Nelson: Edinburgh, 1953) suggests Locke's chapter against innate ideas is mainly directed against the Cambridge Platonists who also make this conflation. p. 59.
5. *An Essay Concerning Human Understanding*, Vol. 1, p. 27.
6. *Ibid.*, Vol.2, p. 11.
7. *Ibid.*, Vol. 2, p. 167.
8. *Das Erkenntnisproblem, in der Philosophie und Wissenschaft der neueren Zeit*, (Darmstadt: Wissenschaftliche Buchgesellschaft, 1973), Vol. 2, p. 260.
9. *Leibniz Philosophical Papers and Letters*, p. 656.
10. *New Essays on the Human Understanding*, (Cambridge: Uni. Press, 1981), tr. Peter Remnant and Jonathan Bennett, p. 52.
11. *Ibid.*, Bk. 3, esp. Ch. 5.
12. *Ibid.*, p. 228.
13. *Ibid.*, p. 233 ff. Also see *The Monadology*. Nicholas Jolley in *Leibniz and Locke: A Study of the New Essays on Human Understanding*, (Oxford: Clarendon, 1984) argues that from Leibniz's side 'the central issue is not epistemological at all; it is metaphysical. The chief focus of Leibniz's hostility to Locke's philosophy is what he takes to be its pervasive materialist tendency. For all its apparent randomness and lack of direction, the *New Essays on Human Understanding*

is a book defending the idea of a simple immaterial and naturally immortal soul.' pp. 6-7. While not wishing to dispute this, it is important to emphasize that neither Leibniz nor Locke makes a clear cut distinction between epistemology and ontology. For both the problems in the one field lead naturally into problems in the other.
14. Kant's orientation is well summed up in the *Prolegomena*: 'High towers, and metaphysically tall men...round both of which there is commonly a lot of wind, are not for me. My place is the fruitful *bathos* of experience, p. 144.
15. *Critique of Pure Reason*, tr. Norman Kemp Smith, (New York: St. Martin's Press), B. 128. All references to the first *Critique* will be to this edition. Hereinafter this shall be abbreviated to *K.r.V.*
16. The extent of the debt pervades the entire undertaking and is mentioned in several places. The most widely referred to acknowledgment is in *Prolegomena*, pp. 115-119.
17. *Enquiries Concerning Human Understanding and Concerning the Principles of Morals.*, (Oxford: Clarendon, 1975), ed. P.H. Nidditch, sect. IV, pt. II, *A Treatise on Human Nature* and *Treatise*, (Oxford: Clarendon, 1978), bk. 1, sect. XIV.
18. *K.r.V.*, B XIII-XIV.
19. This is fundamental to Kant's account of the 'Cosmological Ideas.' See esp. *K.r.V.*, B 446-448.
20. Hume 'was therefore in error in inferring from the contingency of our determination *in accordance with the law* the contingency of the *law* itself.' *K.r.V.*, B 794. See also 'Of the Right of Pure Reason to an Extension in Its Practical Use which is not Possible in its Speculative Use', pt. 1, bk. 1, ch.1 sect. 2, *Critique of Practical Reason'*, (hereinafter *K.p.V*), tr. Lewis White Beck, (Indianapolis: Bobbs-Merrill, 1956), pp. 52-59. All further references to *Critique of Practical Reason* are to this edition.
21. *K.r.V.*, B 146, B 165.
22. For Kant, Leibniz had 'intellectualized the phenomena', while Locke had 'sensified' his concepts of reflection. *K.r.V.*, B 327. Descartes is not mentioned in this respect. But from various other passages, it is clear that Kant does not see his own work as carrying on from Descartes. For Kant, Descartes as well as being a physicist is a 'sceptical' or 'problematic' idealist, whose legacy was not the distinction between the epistemic rules of the mind and sense data, but the 'scandalous' problem of the existence of the external world. *K.r.V.*, B XL, B 274-275, *Prolegomena*, p. 146
23. *K.r.V.*, B 20. Also *Prolegomena*, p. 35.
24. If we do not bear the above in mind, Kant's 'Transcendental Aesthetic' and the 'Axioms of Intuition' and 'Anticipation of Perception' (which are built upon the 'Transcendental

Aesthetic') become incomprehensible. The 'Axioms of Intuition' and 'Anticipation of Perception' provide the real answer for Kant's initial assertion (*K.r.V*, B 14-17) that mathematical judgments are synthetic and not analytic, i.e. mathematics, although an axiomatic science, is not essentially a logical science composed of explicative judgments (*Erläuterungsurteile*), but a constructive science, composed of ampliative judgments (*Erweiterungsurteile*) having empirical application.

25. *K.r.V.*, B 198-202.
26. Kant himself in the 'Transcendental Analytic' indicates that his 'Analytic of the Understanding' is to take the place of 'the proud name of ontology' which claims to have 'synthetic knowledge *a priori* of things in general.' *K.r.V.*, B 303. Kant's 'Analytic' restricts itself to defining the forms, rules and principles which phenomena must conform to. In *What Real Progress has Metaphysics made in Germany since the Time of Leibniz and Wolff?*, tr. Ted Humphrey, (New York: Abaris, 1983), p. 1983), Kant classifies ontology as a part of metaphysics, which deals with 'concepts of the understanding and principles' only in so far as they are related to objects of experience. p. 53. See also *K.r.V.*, B 873-875. His 'transcendental philosophy' is here explicitly equated with ontology, but the ontology is restricted to the conditions and 'first elements' of our *a priori* knowledge. In other words the problem of what is is inseparable for Kant from the problem of the conditions of our knowing.

The debate over whether Kant is an ontologist or epistemologist goes back a long way. In the reaction to the metaphysical and psychological picture of Kant which had been generated mainly by Schopenhauer, a new wave of Kant scholars, who included Kuno Fischer, Hermann Cohen and Alois Riehl came to emphasize and clarify the relationship between the *Critique of Pure Reason* and epistemological foundations for scientific inquiry. (Cohen was also particularly interested in the ethical dimension of Kant's thought, and Fischer's two volume work *Immanuel Kant und seine Lehre*, (Carl Winter, 1898, [1868]) examined almost the entire topography of Kant's thought.) In rediscovering Kant, much space had to be devoted to rectifying psychological and ontological disorientations. In a provocative anti-ontological reading Riehl declared toward the end of the first volume of *Der Philosophische Kritizismus: Geschichte und System*, 'The question of the critique of reason is an epistemological question (eine Frage nach dem Erkenntnis); it is not an inquiry into existence. Its question is neither metaphysical nor psychological; it is critical. It does not want to demonstrate the existence of things, nor ground their essence.' (Leipzig: Wilhelm

Ingeham, 1908 [1876]), Vol. 1, p. 579, my translation. Much later Martin Heidegger began another reading of Kant which is still highly influential in Germany. Heidegger defined his reading against the epistemologically orientated readings of Cohen and the Marburg school. In several works Heidegger emphasizes the ontological and ultimately existential motivations governing the *Critique*. One of the reviewers of Heidegger's best known work on Kant, *Kant and the Problem of Metaphysics*, was a student of Cohen, Ernst Cassirer. Cassirer's critique hit the weak point of Heidegger's reading. Heidegger had located the centre of *Critique of Pure Reason* not only in 'The Transcendental Analytic', but in the faculty of the 'transcendental imagination', which we are to look at later. The importance of this faculty was not in dispute, but Cassirer pointed out that Heidegger had failed to give due care to the necessary dualism required for Kant's moral philosophy. Kant's moral philosophy is not grounded in the 'transcendental imagination' but in the 'typic'. These technical differences imply the difference between the problem of being and temporality (which, of course, is the problem of Heidegger's early work) and Kant's problem 'of is and ought (von "Sein" und "Sollen") of experience and idea (von "Erfahrung" und "Idee").' 'Kant und das Problem der Metaphysik, Bemerkungen zu Heidegger's Kant-Interpretation', *Kant Studien*, Vol. 36, 1931, pp. 1-26. See esp. p. 16. For an interesting debate between Cassirer and Heidegger which throws much light on their different interests in Kant, see *Philosophy and Phenomenological Research*, 1964, pp. 208-222. In Germany Gerald Prauss is now perhaps the best known exponent of Kant as a teacher of a theory of knowledge. But his works place little emphasis on Newtonian science in Kant's thought. The problem of ontology versus epistemology is also prevalent in English works on Kant. Strawson's *The Bounds of Sense: An Essay on Kant's Critique of Pure Reason* is overtly ontological in orientation. In a more recent examination of Kant, Henry E. Allison contrasts the concept of 'an epistemic condition' with an 'ontological condition', making the former the guiding thread of his exposition. See *Kant's Transcendental Idealism: An interpretation and Defense*, (New Haven: Yale Uni. Press, 1983). The distinction is drawn at pp. 10-13. My reading of Kant has profited most from the readings of Riehl, Fischer, Cohen, and Cassirer, who by no means concur on all major points

27. *K.r.V.*, B 395. 'Metaphysics has as the proper object of its enquiries three ideas only: *God, freedom, and immortality.*'
28. See the 'General Scholium' from the end of Book III, added to the second edition of *Principia*. Also compare 'Query 28'

and 'Query 31' of the *Opticks*. These passages are included in the 'Appendix' to *The Leibniz-Clarke Correspondence*, ed. H. G. Alexander, (Manchester: University Press, 1956). See pp. 166-169, 173-174, 180-183. The debate between Leibniz and Newton cannot be underestimated in Kant's problem or his strategy.

29. *Enquiries*, p. 165. Theology, for Hume, is then restricted to revelation (which is open to the criticisms laid against miracles, *Enquiries*, pp. 86-101) and to reasonings of 'particular' and 'general facts'. By means of this move Hume provides the way for an anthropological dissection of religious ideas. Kant's metaphysics, on the other hand, deals only with concepts which cannot be adequately classified by empirical categories. See Kant's discussion of the *Dialogues Concerning Natural Religion* in *Prolegomena*, para. 58.

30. *Prolegomena*, p. 35. These last two questions are the last two 'main transcendental questions' of *Critique of Pure Reason*. At *K.r.V.*, B 22, Kant asks '*How is metaphysics as natural disposition possible?*'

31. *K.r.V.*, B 59. The justification of such a metaphysic for Kant ultimately resides in space and time being *a priori*. We have yet to consider why Kant held this revolutionary, though now defunct view, which is so decisive for the strategy and content of Kant's entire philosophy.

32. On the importance of genius, *The Critique of Judgement*, (hereinafter *K.d.U.*), tr. James Meredith, (Oxford: Clarendon, 1952), pt. 1, para. 46-50.

33. *K.r.V.*, B 1. Because Kant solves his problem with cognitive faculties, it is easy to transform his problem into a psychological problem. For the critique of Kant as undertaking a psychological problem see Riehl *op.cit.*, p. 363 ff. and Hermann Cohen, *Kant's Theories der Erfahrung*, (Berlin: Bruno Cassirer, 1918 [1871]), pp. 254-310. One of the most widely read commentaries in English on Kant is Norman Kemp Smith's *A Commentary to Kant's Critique of Pure Reason*, 2nd. ed., (London: 1923). Kemp Smith takes issue with Riehl and Cohen and resurrects the psychological reading of Kant. See pp. 51, 102. The price he pays for this move is the failure to find coherence in either the problems or execution of Kant's philosophy.

34. *K.r.V.*, B X-XI.

35. *K.r.V.*, A XIII. Kant then continues with the claim that pure reason is a complete unity. The reason he held this view becomes apparent as we consider the relationship between the metaphysical foundations and the development of metaphysical principles. To understand this relationship requires understanding why metaphysical foundations must be synthetic judgments.

36. *K.r.V.*, B 10, 11.
37. *K.r.V*, B 191.
38. *K.r.V.*, B 11-12. *Prolegomena*, Sect. 2.
39. Cassirer, *Das Erkenntnisproblem*, Vol. 2. pp. 679-680 points out that weight was considered by almost the entire scientific community to be an empirical and not an essential attribute of matter. There are several discussions on the nature of analytic and synthetic judgments. Riehl throws much light on Kant's specific examples, *op.cit.*, pp. 419 ff. Also see Lewis White Beck 'Can Kant's Synthetic Judgments be made Analytic?' in *Studies in the Philosophy of Kant*, (Indianapolis: Bobbs-Merrill Co., 1965); Allison, *op.cit.*, pp. 78-80 is also useful.
40. *K.r.V.*, B 17. This judgment is later to be classified by Kant as the first 'Analogy' of Experience.
41. If we fail to realize this the 'Analogy' makes no sense whatever. See Carl Friedrich von Weizäcker's brilliant exposition of the first 'Analogy', 'Kant's "Erste Analogie der Erfahrung" und die Erhaltungssätze' in *Kant: Zur Deutung seiner Theorie von Erkenntnis und Handeln*, ed. G. Prauss, (Köln: Kiepenheuer, 1973), pp. 155-163.
42. Kant's own assertion about this judgment in the 'Introduction' is highly misleading. The 'proof' that the conservation principle is synthetic and *a priori* is supplied in the first 'Analogy'. That proof, like the proofs of all the synthetic *a priori* judgments, requires a grasp of the *a priori* elements which Kant gradually introduces throughout the course of the *Critique of Pure Reason*, and a grasp of how he arrives at precisely these and no other elements. In the 'Introduction' Kant presents the results, but it may appear as if they are the assumptions of *Critique of Pure Reason*.
43. *K.r.V.*, B. 19.
44. Immanuel Kant, *Logic*, tr. Robert S. Hartman and Wolfgang Schwarz, (Indianapolis: Bobbs-Merrill, 1974), para. 36.
45. Hence one of Kant's best known sentences: 'Thoughts without content are empty, intuitions (*Anschauungen*) without concepts are blind.' *K.r.V.*, B 75.

2 The strategy and apparatus of Kant's critical philosophy

The very problem of specifying which kinds of knowledge (*Erkenntnisse*) are synthetic *a priori* was to be the key to answering the other problems we have discussed. For if there are elements and principles which are not derived from experience but are its constitutive and regulative conditions, the logical question is to ask whence do these elements and principles derive? Kant answers that they must derive from the faculties or abilities (*Vermögen*) that are (a) peculiar to the human mind and (b) indispensable to the formation of a judgment that purports to be objective. The task then was to isolate the faculties which are used in the act of knowing and then specify what elements, if any, were essential to the very operation of the particular cognitive faculty under inspection. If any elements could be shown to be necessary formal conditions of the very operation of the specific cognitive act, then they would serve as the formal conditions of any empirical data to be cognized. In other words they would be necessary and *a priori* conditions of any objective judgment of experience.

If one could find these elements then one had to inquire whether and how these elements could form a unity. That is to say the next step was to demonstrate how the *a priori* elements could be *synthesized* into *a priori* principles. Just as the specific *a priori* elements had to be traced to a necessary cognitive source, the principles themselves also had to be traced to a cognitive source. Thus we see that Kant's 'scientific metaphysic' required locating the source of the *a priori* elements and principles. The location of the source, i.e. the demonstration of the site and necessity of specific *a priori* elements of cognition, was to make possible the specification of the legitimate scope of the elements. Because the elements were *a priori* and because their source had never been specified, it was, Kant holds, previously quite natural to employ them without paying due care to their legitimate scope.

The three formal operations that Kant held to be indispensable for theoretical knowledge were: (1) the intuition (*Anschauung*) of sense data; (2) the combination of sense data into judgments; and (3) the logical inferences which co-ordinate different judgments. Stated otherwise, Kant held that knowledge is a process involving: (1) receptivity of sense data; (2) the classification of that data and the combination of the classifiers in judgments; and (3) the systematic co-ordination of the judgments by inferences.[1]

These three operations are classified by Kant as the faculties of sensibility, understanding and reason. The *a priori* elements and principles, which Kant claims to have discovered after isolating these faculties, are: (1) the forms of intuition; (2) the functions of the understanding, and when these are synthesized with the forms of intuition, the 'Synthetic Principles of the pure Understanding'; and (3) the ideas and the ideal of reason. The dissection of the faculties involves a 'Transcendental Aesthetic', which is an investigation into the formal and *a priori* conditions of sensation, and a 'Transcendental Logic' which investigates the *a priori* conditions of judgment and the syllogisms.

The transcendental ideality of space and time

The 'transcendental ideality' of space and time is, by Kant's own account, one of the two pivots around which the critical philosophy turns. (The other is the concept of freedom.)[2] Simply put the 'transcendental ideality' of space and time means that the properties of space and time are not objectively perceivable like other material or empirical objects, but that they are 'subjective conditions' of experience.

Central to the conception of space and time that Kant is attempting to explain is the idea in classical mechanics of the potential indefinite divisibility of space, time, and matter. The infinite is not found in experience; it is a mathematical concept. Yet in classical physics this mathematical concept has direct empirical application when phenomena are broken down indefinitely into ever more complex and divisible phenomenal relations, into ever more mechanical motions taking place over ever smaller spaces and for ever shorter durations.

Space and time appear to possess the attributes of infinity. Within the classical framework space is an infinite and infinitely divisible magnitude consisting of homogeneous parts. Likewise, time consists of homogeneous parts; different times are only parts of time. Because of these attributes and because space is a precondition of spaces, and time is a precondition of times, Kant distinguishes the concepts of time and space from all other empirical concepts. Other universals do not contain particulars within them.[3] A dog does not contain real dogs within it. A dog is not divisible into infinitely many dogs. The concept dog is an abstraction from particulars; the universal is derivative from the particulars. But space and time are infinitely divisible into spaces and times, and it would appear as if absolute space is antecedent to the particular space, just as absolute time is antecedent to particular times. (I say seem because Kant's arguments gain their force from a conception of space and time as absolute, but Kant's conclusion is that they are not absolute things; they are only absolute as ideas.)[4]

In sum space and time share fundamental affinities with the mathematical concept, infinity. Yet Kant argues that they have another special characteristic. We do not empirically observe space and time, nor their properties. We do not perceive the continuity, homogeneity, immutability and infinitude of space. Rather we perceive empirical things. Yet we can construct mathematical/mechanistic models based on assumptions that space possesses these attributes, and this in turn enables us to formulate mechanical laws. Likewise, we do not perceive the properties of time, co-existence, and succession. We only perceive things co-existing or succeeding each other. We perceive things in time, not time, just as we perceive things in space, but we never perceive space. If we do not perceive space and time but only things in space and time, how do we know their properties? To answer this requires closer observation of the cognitive faculty which Kant isolates in the 'Transcendental Aesthetic'.

This is the faculty of viewing or intuiting things, *Anschauung*. If we can only perceive or intuit things if they are somewhere at sometime, then time and space are *a priori* conditions of an object being viewed. Kant disqualifies as empirically real whatever could never be representable in space or time. To this extent space and time are seen by Kant as *a priori* conditions of empirical intuition. It is also possible to make a clear distinction between empirical attributes which accompany sensation, i.e. hardness, colour etc., and the attributes of extension and shape. The latter are mathematical constructions and are the work of the mind. On the basis of this distinction Kant separates pure intuition (*reine Anschauung*), the capacity of the mind to view or intuit its own mathematical constructs, from empirical intuition, i.e. the capacity to view empirical objects.[5] This separation leads Kant to make the formal conditions of pure intuition, i.e. the conditions of mathematical and mechanistic constructions, the conditions of empirical viewing. Space and time are the *a priori* conditions of both types of viewing, the viewing of mathematical/mechanistic constructions and empirical objects. Time and space thus, for Kant, lay at the foundations of mathematics, and provide his answer to how mathematics is applicable to experience. This needs to be explained more fully, for if we fail to appreciate what Kant is doing here, his philosophy makes little or no sense.

Firstly, time is for Kant at the basis of the most fundamental mathematical operation, the aggregation of units. Kant has in mind the action of counting, and for mechanics, this becomes synonymous with the initial construction of the number line and the axes of co-ordinates. The central idea of Kant's remarks about mathematics in *Critique of Pure Reason* and the *Prolegomena*, that the foundation of mathematics is in the production and reproduction of the unit, relies upon the idea that integers are possible because of the capacity to synthesize units. In *Attempt to Introduce the Concept of Negative Numbers into Worldly Wisdom* Kant had argued that subtraction is merely a directional operation, which itself requires the performance of the same ability to accumulate units as addition does.

> And because subtraction is a cancellation which happens when contrary numbers are taken together so it is clear that the sign '-' is not actually a sign of subtraction as it is usually represented, rather the '+' and '-' together only describe a transference (*Abziehung*). Hence '-4-5=9' was no subtraction, but it was a genuine increase and combination of numbers of the one kind. However, '+9-5=4' means a transference, in so far as the sign of opposition indicates that the one cancels the other when they are equal.[6]

In other words the primary arithmetical operations of addition, subtraction, multiplication, and division are dependent upon the act of the synthesis of units. This act of synthesis is at the basis of Kant's claim that mathematical judgments are not based upon the purely logical law of contradiction. They are not analytic, they are synthetic. Essential to this claim is that each cardinal number is an aggregation of one or more units. Kant takes the judgment 7+5=12 and he claims that it is synthetic because the answer is not reached by an analysis of the numbers seven and five, but by the synthesis of the numbers. To make his point Kant talks of resorting to intuition, of counting one's fingers in order to reach the total. Because of the simplicity of this particular example one is able to think the answer automatically. And thus Kant says that the idea becomes clearer when larger numbers are considered. Kant points out that mathematical judgments are usually considered to be analytic because, as is evident in this example, there is an answer that should be thought. But Kant claims that this is not relevant.[7] What is relevant is that the operation requires an act by the subject which extends beyond merely analysing the numbers given. The science of number is not, for Kant, explicative but ampliative. It initially requires the construction of its units. Reflection upon the number 1 does not necessarily lead me to any other number. As Kant pointed out in a letter to Schultz, when I think of a number I do not consider all possible combinations which could result in this number. But, if the judgment were analytic, this would be so. The judgment 4+3=7, says Kant, is not a variant on the judgment 12-5=7, as if an analysis of the former judgment would lead me to consider the latter one.[8]

Ernst Cassirer has described well the thrust of Kant's conception of mathematics as synthetic and constructive.

> What Kant meant is that the fundamental character of the synthesis at which mathematics aims comes to light not so much in the formation of concepts and judgments as in the building up of the mathematical world of objects. The formation of the objects of mathematics is constructive, and hence synthetic, because it is not concerned simply with analyzing a given concept into its characteristics, but because, starting from certain determinate basic relations, we advance and ascend to ever more complex ones, where we let each new totality of relations correspond to a new realm of "objects."
>
> The development of the realm of numbers fully confirms *this* meaning, which Kant associated with the idea of mathematical synthesis. There can be no question that the concept of the whole number includes all those features from which the notions of rational, irrational, and complex numbers were later derived. Such essential stipulations of the concept as this, that every whole number has an immediate predecessor and successor, had to be abandoned in order to achieve the other number systems.[9]

In *Critique of Pure Reason* Kant is attempting to establish the condition of the possibility of mathematics. Once he can establish what type of judgments mathematics are he can then define their scope. If, as Kant argues, arithmetical judgments are based on an act of synthesis, and if that synthesis initially takes place over time, then Kant is able to create a barrier against the Platonists and other 'enthusiasts' (*Schwärmer*)[10] who use mathematics as the point of entry into the non-sensuous realm. Having 'discovered' a purely rational realm they engage in all sorts of speculative claims about purely intelligible objects. This way is blocked if numbers no longer reside in a purely intelligible realm, but instead are the product of a mind which can initially only aggregate units over time. Mathematical entities have no existence apart from time. The rationalists have been correct in pointing out that they are not empirical, but wrong in divorcing them from sensation. Neither empiricists nor rationalists have made construction, under the condition of succession, the key to the foundations of arithmetic. (To repeat, Kant is not speaking of historical foundations, but about how a mathematical system is developed axiomatically.) Concomitantly neither rationalists nor empiricists have explained how number can be applied to empirical data to an infinitesimal degree, as happens in mechanics.[11]

Having noted the fact that mechanics bridges the 'two worlds', the mathematical and the empirical, Kant has sought to explain how it does so. He has concentrated on the fundamental act of arithmetic, the production and reproduction of the unit, and made that act inseparable from the concept of time. Time is the middle term between the pure science of mathematics and physics, the *res media* between a pure science of the mind and a science of experience. The connection is vital to the 'Axiom of Intuition' and is succinctly brought out in paragraph 10 of the *Prolegomena* where Kant writes:

> Arithmetic forms its own concepts of numbers by successive addition of units in time; and pure mechanics especially can only form its own concepts of motion by means of the representation of time.[12]

In light of the above considerations we can now make sense of Kant's claim that time is transcendentally ideal. Time is transcendentally ideal because it is a condition of pure operations of intuition as well as any empirical data. Time is not something apart from either the empirical or the mathematical objects which can be represented in empirical or pure intuition. It is not something in-itself. It is not an absolute in this sense. It has neither function nor meaning apart from the ideal and empirical data which can be constructed and received by the mind of the subject. To this extent Kant calls time a 'subjective' condition, a mere form of experience.[13] The subjectivity is not to be taken psychologically, as if Kant sees the development of the world (as opposed to our *judgment* of it) as dependent upon our consciousness of it. Moreover, psychology like every empirical science is itself to be studied under the condition of time.[14] Time, for Kant, is a universal condition of experience. But, to repeat, because it is the condition of the construction of the unit and the infinite divisibility of a unit as well as any phenomenon, it is a pure form of intuition. As a pure form of intuition Kant also claims that it is empirically real. What he means by this is that its properties of succession and co-existence can only be known by reference to co-existing and successive things.[15] But because mathematical relations can be successive and co-existent, time is not an empirical 'thing' alongside other things, it is the condition of phenomena, and of the aggregation of units.

Time, however, is only one form of intuition, and we have implicitly relied upon the other form, space, to clarify what Kant is doing. For the construction of successive units requires the objectivication of that construction, and space provides the ground for that objectivication; by being the condition for the construction of a number line.[16]

While time is the precondition of the aggregation of units, space is the precondition of the construction of the figure. As in the case of arithmetic, Kant argues that the truth of a geometrical judgment is a matter of the construction and amplification of one's concepts, not merely an analysis. Kant gives a simple illustration of what he means by geometrical construction in the 'Doctrine of Method' where he compares the activity of a geometer with a philosopher. If a philosopher, unskilled in geometry, is given a triangle and asked to demonstrate the relationship that its angles have to a right angle the philosopher will not possess any special advantage in solving this problem. If the concepts of the triangle are broken down analytically so that the subject 'triangle' has the predicate 'three lines and three angles', we are not a step closer to the concept of two right angles. The geometer on the other hands makes the demonstration by the construction of concepts.

> He at once begins by constructing a triangle. Since he knows that the sum of two right angles is exactly equal to the sum of all the adjacent angles which can be constructed from a single point on a straight line, he prolongs one side of his triangle and obtains two adjacent angles, which together are equal to two right angles. He then divides the external angle by drawing a line parallel to the opposite side of the triangle, and observes that he has thus obtained an external adjacent angle which is equal to an internal angle — and so on. In this fashion, through a chain of inferences guided throughout by intuition, he arrives at a fully evident and universally valid solution of the problem.[17]

As this example illustrates, geometry is able to be conceived as consisting of synthetic judgments which are *a priori* because the geometer must perform an act of synthesis, and that act, for Kant, takes place only under the condition that it comply with spatial conditions. Hence Kant concludes that space is a condition of the construction of figure, just as time was for the construction of number. The conditions are not the result of a logical law, rather they are only revealed through the act of construction of the figure. The qualities of Euclidean space are therefore construed by Kant as dependent not upon empirical experience (how could the properties of homogeneity, infinity, continuity and immutability be perceived in experience?), but upon the form under which the construction takes place.[18] This is the form of the intuition or the viewing of the construction.

Just as time was both a condition of an ideal series, and of sense data being construed as an extended magnitude, space is also a condition of that construction as well as of sense data in so far as sensations can be represented geometrically. Once again the conditions of the ideal construction are seen by Kant as the same conditions as those of experience. Once again a middle term is found between empirical data and the ideal and axiomatic relations of the mind. Hence Kant concludes that space is also an *a priori* condition of intuition, for it is a condition of pure intuition, i.e. a condition of the construction of pure geometrical relations, and empirical intuition. He holds that space, like time, is transcendentally ideal. And this means that space, like time, cannot be considered as an absolute in itself, as Newton is criticized for believing. It has no meaning or function apart from the act of intuition, i.e. it has no meaning apart from the empirical data (which, if it exists, must occupy some space), nor apart from the ideal geometrical constructs. If isolated from the objects of pure or empirical intuition, it is nothing. It becomes, as Kant believed it did in Newton's theological speculations, the subject of fantastical and absurd speculations.[19] Absolute space is a mere idea. What we have are kinematic frameworks which we employ for particular descriptions of motion. As Kant writes in *Metaphysical Foundations of Natural Science*:

> absolute space is in itself nothing and is no object at all, but signifies merely every other relative space that I can at any time think of outside a given space, and that I merely can extend beyond each given space to infinity as being such a space as includes this given one, and in which I can assume this given one, and in which I can assume this given one to be moved.[20]

In making this move Kant is not sacrificing the classical idea of infinite divisibility of spaces or of times. He is, however, excluding the legitimacy of any metaphysical proclamations about concepts (absolute time and space). Hitherto the great opponent of Newton's conception of absolute space and time had been Leibniz.[21] For Leibniz space and time are not independent things that can be without bodies, they are relative. Space is an order of co-existences, and time an order of succession.

Because of the pervasive presence of Leibniz in Kant's 'Transcendental Aesthetic', and because Kant is steering a course (as he invariably does when reconciling metaphysical disputes) between two opposing positions — in this case Newton and Leibniz — it is necessary to pause upon Kant's critique of Leibniz's conception of space and time. Kant's conception of space and time coincides with Leibniz in so far as both agree that spaces and times are nothing without the empirical or constructed relations that appear in space and time. But, according to Kant, the price Leibniz pays in order to acquire his conception of relative space and time is a conflation between experience and logic. The problem becomes particularly visible, for Kant, in the implication Leibniz's thought has for the nature of geometry. Kant holds that Leibniz and his followers make the *apodeictic* rules of geometry as contingent as the relations of experience. Speaking of the Leibnizians in *The Inaugral Dissertation* Kant says:

> they throw down geometry from the summit of certitude and thrust it back into the rank of those sciences whose principles are empirical. For if all the affections of space are merely borrowed by experience from external relations, there is only a comparative universality present in the axioms of geometry, of the kind that is obtained by induction, that is, extending as far as it is observed. Nor is there present any necessity except what is arbitrarily constructed, and there is hope, as happens with things empirical, of uncovering sometime a space endowed with different primitive affections and perhaps even a rectlinear figure enclosed by two straight lines.[22]

In the *Prolegomena* Kant extended his critique of Leibniz's definition of space. Leibniz, claimed Kant, fails to account for the fact that identical geometrical properties can be represented by figures that are nevertheless not congruent. Kant gives the example of two spherical triangles, with identical sides and angles, on opposite hemispheres which have an arc of the equator as their common base. He also gives the example of a hand and its mirror-image.[23] In both cases, argues Kant, we must rely on nothing other than our view, i.e. our intuition, of them to clarify the difference in relations between these figures. Leibniz's problem is that he does not distinguish between things as they are in themselves and things in so far as they conform to the forms of intuition, space and time. Indeed, Kant believed that Leibniz had really been groping for the doctrine that Kant himself had discovered.[24] For, in Kant's doctrine of the ideality of space and time, Leibniz's conception of relative spaces and times remain untouched, but the conflation of mathematical points, i.e. points outside of space and time, with matter is avoided, as is the conflation of geometrical truths with inductive truths.25

Moreover, Kant claims that the contradiction between Leibniz's monad, the mathematical absolutely simple, and the physicists's postulate of the infinite divisibility of matter rests upon the mistaken belief that space and time possess the same ontological status as things. Once space and time are seen as forms of intuition, forms with which empirical sensations and sensory constructs *must* conform, the doctrine of infinite divisibility in space and time remains, for Kant, because space and time must have some content (whether empirical or mathematical) for them *to be*. The doctrine of absolute simples must refer to something outside of space and time. Because an absolute simple cannot be found in time and space, it, like Newton's absolute space and time must refer to a 'mere idea', a heuristic postulate of reason.26

In effect Leibniz and Newton represent for Kant two opposing sides of the one failure, a failure Kant attributes to all rationalists and empiricists. We must now focus upon the nature of that failure.

The significance of the distinction between appearances (*Erscheinungen*) and noumena in Kant's strategy

The failure of Newton and Leibniz, and indeed the failure that had hitherto been the source of endless metaphysical disputes, for Kant, is the failure to distinguish adequately between things which conform to the forms of intuition, (appearances) *Erscheinungen* and things in themselves (noumena). The 'transcendental ideality' of space and time is the first step in a work whose results largely revolve around this distinction. *Erscheinung* or appearance is for Kant the material of an empirical intuition. As he states early in the 'Transcendental Aesthetic', 'The undetermined object of an empirical intuition is entitled *appearance* (*Erscheinung*).'27 Given that space and time are the conditions of empirical intuition, an appearance is something that is somewhere at sometime.

Both Leibniz and Newton use the distinction between things in themselves and appearances, and the idea is clearly in Locke.28 The epistemological sense of the distinction derives from the finitude of our perceptual faculties. Because we must acquire knowledge through our sense organs we are restricted in what we can know. We only ever observe a restricted range of relations covering a restricted time span and a restricted spatial extension. We perceive only some of the mechanisms and relations which contribute to the formation of sense data. In this sense we observe the appearances of things. An intelligence that could observe in one intuition all relations stretching forwards and backwards in time, and in all spatial directions, would not be restricted to viewing the appearances. Such an intelligence would know the things themselves. And such an intelligence is, for Leibniz and Newton, God. This distinction relies upon two different types of intelligences: one is finite, it builds up knowledge over time by piecing together the relations it is able to discern (through experiment, and trial and error); the other is unlimited or infinite.

But while this distinction between noumena and appearances was common enough, Kant disputed the way it had traditionally been used. Hitherto the distinction had been one of degrees, as if an infinite intelligence were the perfection of a finite intelligence.

This is most evident in Leibniz's thought. For Leibniz intelligences with sensory organs have a limited perception of the infinite relations or compounds into which it enters. God differs from other intelligences in that His ideas 'are infinitely more perfect and extensive than ours.'[29] Or to use a standard maxim of Leibniz, God knows all possible events in all possible worlds. While God sees clear perceptions, because He knows His plan, when we observe sensible qualities our perceptions are 'blurred.'[30] However, according to Leibniz, through the study of the clear and intelligible relations, i.e. eternal and necessary ideas, we can know God's plan. In other words we can know part of the mind of God. And the dispute with Newton, in which Clarke served as an intermediary, is not just about the nature of space and time, but about the mind of God, and God's plan. The dispute about physics has a metaphysical core.

Within this dispute neither party can provide anything other than speculative arguments for his position. There is no empirical way of testing what they are saying. But the metaphysical dispute, which rests upon things in themselves (i.e. the ideas of God), is not only a dispute in which questions of physics and metaphysics are interlocked. In Leibniz's case his ostensible knowledge about the ideas of God is directly connected to considerations on justice and morals generally. For Leibniz, the intelligible world is synonymous with God's moral kingdom, the kingdom of grace.[31] It is this move which makes him, rather than Newton, or for that matter any one else with the possible exception of Hume, Kant's major opponent in the *Critique of Pure Reason.*[32]

Leibniz's conception of 'the moral kingdom of grace' stands in close relation to two of his other doctrines which play a major role in the tactics adopted by Kant in *Critique of Pure Reason*: the one is that noumenal ideas are merely clear representations of perceptions; the other, and this flows from the first, is that our world is the best of all possible moral worlds. The first idea indicates that moral ideas differ from phenomenal ones by degrees, not by kind. The second indicates that the world is guided by providence. All existents are ultimately the work of a divine plan. For Kant, Leibniz thus has no room for the autonomy that is essential to genuine moral freedom. (Hence in *Critique of Practical Reason* Leibniz is said to be an advocate of a kind of determinism: he is a spiritual and psychological determinist.)[33] For Kant this results directly from Leibniz's failure to distinguish adequately between noumena or things in themselves and phenomena. Leibniz makes a logical distinction, but he fails to see that there are two different types that can be isolated and traced to their cognitive sources. In other words he failed to make an adequate distinction because he had not undertaken a *Critique of Pure Reason*. (This is why Kant was so annoyed with Eberhard's claim that Leibniz had already undertaken such a Critique.)

But it is not only the dispute between Leibniz and Newton which rests upon this failure to distinguish precisely between judgments referring to supersensuous or noumenal ideas and those referring to phenomena.[34] What, for Hume, is not derived from fact or mathematics is fantasy. The connection between metaphysical rationalism and metaphysical scepticism is, for Kant, a necessary one. They both result from the absence of a scientific metaphysic.[35]

Rationalism may focus upon principles that are heuristically useful and perhaps even necessary for science, but only continued testing of a theory against the phenomena can lead to the progress of science. Although the project of Kant takes as its starting point the need to specify which *a priori* elements and principles are necessary for mathematical science, it is not of primary importance for the advancement of science that scientists recognize the metaphysical nature of the foundations of science. Their job is to apply and discover empirical rules, and if there are necessary *a priori* conditions which make that activity possible then they will be forced to assume them. In other words, if, as Kant claims, objective judgments of experience are governed by synthetic principles of pure reason (such as the law of cause and effect, the reciprocity of material relations etc.), then the mathematical physicists will not be able to escape metaphysics if they are to make headway.[36] At the same time Kant did see scepticism as striving to destroy the foundations of knowledge, and the status of science. In this sense it could easily ally itself with the forces of anti-Enlightenment. And thus it had to be countered. But Kant distinguishes between scepticism and the sceptical method. The latter, by aiming at certainty, undermines groundless beliefs. Kant integrates the sceptical method into the critical philosophy in his fight against metaphysical dogmatism.[37]

Nevertheless, there is, for Kant, also a dogmatic side to the strict empiricism employed by the sceptic. Strict empiricists demand that human judgments refer to *what is*. From this position what ought to be must ultimately stem from some judgment of fact, whether it be about human nature, common sentiment, human inclination, or custom.[38] Thus, for Kant, in place of the invisible, untouchable ideas of moral virtue, which he sees as necessary for the elevation of human beings, the empiricist degrades humanity by making inclinations the criteria of judgment.[39] Morality, for Kant, is concerned not with facts but with unconditioned duties. The idea of moral freedom is not representable empirically or mathematically. It consequently does not conform to the 'transcendental ideality' of space and time. It is thus not an appearance. It is noumenal.

The problem of the legitimation of noumenal ideas, however, cannot be answered until Kant has circumscribed the limits, i.e. the *a priori* conditions of judgments of facts. The 'transcendental ideality' of space and time has meant that no noumena can be found in space or time. 'properties that belong to things in themselves can never be given to us through the senses.'[40] Yet scientific discourses, as we have seen, have crossed over into the field of noumena. Instead of siding with the sceptics and strict empiricists by prohibiting entrance to that field, Kant has to specify why that crossing takes place in science, and then why in morals this field is legitimate as well as valuable. To do this requires probing the faculties of concepts and ideas, the faculty to comprehend or judge (*Verstand*) and the faculty to infer (*Vernunft*).

Transcendental logic part 1: The acquisition of a priori concepts

From the outset of the *Critique of Pure Reason* Kant's philosophy has assumed that knowledge involves a dual process: (1) the receptivity and construction of objects (the *a priori* conditions of this process are presented in the 'Transcendental Aesthetic') and (2) the comprehension or classification of the objects. These cognitive acts work with two different types of representations (*Vorstellungen*): objects of intuition and concepts.

If we could know things simply by viewing them without relying upon the store of past experiences which are accumulated and connected in our concepts, we would have an 'intuitive understanding.' For an intuitive understanding there would be no difference between classifiers (universals) and the particular objects themselves. (Note that, for Kant, an object is any perceivable composite of relations — despite some claims to the contrary there is no technical difference between *Gegenstand* and *Objekt*. Hence an indefinite empirical object is called a manifold by Kant.) An infinite intelligence would not need science, for it would know all relations in their immediacy.[41] Human intelligence, however, cannot know all things at once. There is a discrepancy between universal and particular, which reflects the cognitive dualism of sensation and comprehension. This cognitive dualism defines us as intellectually finite. For us, relations in space and time must be continually observed if they are to be known. Hence we will never have total knowledge. However, because human intelligence is partly governed by sensibility it is able to divide indefinitely the relations it observes into ever smaller spaces and times. Conversely, it can co-ordinate ever more relations over ever broader spaces and greater durations. Our senses are able to discern ever finer differentiae, and our concepts by reflecting these observations become sharpened. Likewise our conceptual relations are broadened by the discovery of greater generic affinities. Concepts are the store of human observation and experience.[42] And taken in their entirety they compose a vast evolving network of mediations. But the assessment of the truth of these concepts will depend upon their correspondence to sense data, or when one is forced to make inferences which are unable to be confirmed by witnessing viewable relations, the concept must not contradict the viewable sense data. Kant's one exception is when the concept is *a priori*.

The 'Transcendental Aesthetic' had argued that there are two *a priori* forms of sensibility. The next step is to see if there are formal *a priori* concepts. To discover them he must isolate the capacity we have to transform the data that is represented immediately as sense data or mediately as concepts into judgments. This is the capacity to judge or understand. Kant makes these terms equivalent, for when we understand something we are making a judgment about what it is.[43] Hence, Kant inquires whether there are any elements which are *a priori* and which are necessary conditions in the act of judgment. Were Kant concerned with specifying the functions of judgments *per se*, his task would be equivalent to that of the logician. But Kant is interested in specifying the synthetic judgments which are *a priori* and necessary for objective judgments of experience, *Erfahrungsurteile*.

To the extent that Kant is inspecting the functions of judgment he is doing a type of logic. But to the extent that this logic is specifically related to objects of experience, Kant is doing something very different from the logician. For logic is indifferent to the material or the source of judgments. Kant's inquiry into 'the functions of understanding' is called a 'transcendental logic.' Unlike general logic it is to be a 'logic of truth'. 'For no knowledge can contradict it without at once losing all content, that is all relation to any object, and therefore all truth.'[44] It is to serve as a 'canon' or yardstick (*Probierstein*) which must be complied with if a judgment purports to be objective.[45] A 'transcendental logic' must set out the formal conditions of judgments of experience.

This is what Kant calls the 'negative service' of the critical philosophy, the safeguarding of the legitimate boundary of scientific or theoretical judgments.[46]

To provide such a yardstick the following tasks have to be fulfilled. Firstly Kant must present the *a priori* elements of the understanding, then he must demonstrate that the act of judgment is not just governed by associations, as Hume claimed, but that there is a necessary form of judgment. This requires an examination of the claim of necessity in the act of cognition. If, as Kant claims, knowledge involves the synthesis of intuitions and concepts then there must be a faculty in which that synthesis occurs, and it must be demonstrated to be a necessary faculty of knowledge. And finally Kant must demonstrate how the *a priori* elements of intuition are co-ordinated with the *a priori* elements of the understanding. The specific co-ordinations will provide the specific synthetic judgments which are *a priori* necessary for a judgment of experience.

These are the steps of the first part of the 'transcendental logic.' They are the vital moments of what Kant calls the 'Transcendental Analytic', the task of which is defined by Kant as:

> the hitherto rarely attempted *dissection of the faculty of the understanding* itself, in order to investigate the possibility of concepts *a priori* by looking for them in the understanding alone, as their birthplace, and by analysing the pure use of this faculty.[47]

For our purposes it is necessary only to clarify the major strategic moves that Kant has undertaken rather than inspect every detail of his claims.

The first step that Kant makes is to draw up a list of logical functions that are to serve as a 'clue' (*Leitfaden*) to the discovery of the primary *a priori* elements of the faculty of understanding. Because the stipulation of the rules of judgment is the task of logicians, Kant uses their work as a clue to discovering the pure elements of the understanding. Then Kant draws up a list of logical functions. But his list of functions necessarily departs in important ways from previous classifications because Kant adds functions which are applicable specifically to objects of a possible experience.[48] In other words the functions of judgment of general logic are tailored for the requirement of a 'transcendental logic'. The list he provides is supposed to be complete. Only thus can it serve as an adequate clue or guide to the necessary elements of the act of understanding. But this claim to completion must be understood in a very specific sense; the list adds logical functions which would not normally be considered distinct by logicians. Thus it may include more logical functions, but it must not omit any.[49]

After Kant has drawn up his list of logical functions and explained his innovations, he then reformulates and reclassifies them as 'the categories' or '*Stammbegriffe*' of the 'understanding.'[50] The sense of this reformulation is very obscure if separated from the remainder of the 'Transcendental Analytic.' Kant has merely attempted to place along side a list of logical classifiers a set of concepts which he will later argue are the *a priori* components in mechanistic science, and in any objective judgment of experience. He leaves it to the reader to see the connection between the elements of a scientific judgment and the logical classifiers. In other words, the legitimacy of the 'categories' of the understanding rests upon their being necessary classifiers of objects of experience. But this legitimacy can only be demonstrated by showing how the particular concepts are co-ordinated with the forms of intuition (space and time), how they relate specifically to viewable objects. Their legitimacy necessarily involves the demonstration of how synthetic judgments are *a priori* possible. The demonstration of this possibility is the main task of the section of the Critique entitled the 'Transcendental Deduction'.[51] The clue to understanding this most difficult and important part of *Critique of Pure Reason* lies in Kant's conception of judgment.

By a judgment Kant means the subsumption of different representations under a common representation. Note that this very definition is based upon the initial dualism in types of representation (intuitions and concepts). The common representation is the concept and that is our contribution. We receive phenomena but *we classify* it. Hence Kant says that concepts depend upon functions; functions are the 'spontaneity' of thought.[52]

Now imagine that we are viewing phenomena without making any type of judgment. No specific relations in space would be distinguished from any others, nor would there be any connections made between relations over time. We would be completely passive recipients of impressions. This isolation of the act of sensation from the act of classification is the crucial separation in Kant's 'Transcendental Deduction'. Indeed we can say that if one fails to grasp this distinction and the specific definition of judgment that has been provided above, the 'Transcendental Deduction' becomes a rambling and chaotic confusion. The 'Transcendental Deduction' is Kant's attempt to stipulate the precise nature and establish the necessity of that spontaneity which occurs when sensory representations are brought together in a judgment. The bringing together (*Verbindung*) of representations is the fundamental act of understanding or judging.[53] Note that this act is not dependent upon any specific object of intuition. Rather it is an act that is required for every object of intuition, and for concepts, which must, if they are to be objectively verifiable, eventually refer to something that is possibly viewable, something occupying space for some time. This act is designated by Kant as a 'synthesis', and he also calls it a self-activity (*Selbsttätigkeit*).[54] Note also that if this act of spontaneity is required for combining any representations, any analytic statements must be originally dependent upon it. (Even *the* analytic science, logic, is said by Kant to be dependent upon it.)[55] As Kant says, 'where the understanding has not previously combined, it cannot dissolve, since only as having been combined *by the understanding* can anything that allows of analysis be given to the faculty of representation.'[56] This act thus supplies for Kant the ground of a possible scientific metaphysic.

If the act of conjoining representations is antecedent to the specific sensuous representations (but not the *Erscheinungen per se*), then it is evident that this unity is not a consequence of the sensuous representations. The importance of this tautology for Kant should not be underestimated. It contains in embryo his refutation of Hume's empiricism as well as rationalism. The unity which Kant calls the 'original synthetic unity of apperception' is the basis of all judgments: empirical, mathematical and pure or *a priori* judgments.[57] It is not a thing, but an act. Kant says it is not even a concept but the necessary accompaniment, the vehicle of all concepts; for it is not like other representations, but their necessary condition of comprehension.[58] This act is, for Kant, the understanding itself. When it is hypostatized and takes on spiritual qualities, it is illegitimately transformed into an object of rational psychology.[59] Note that no representations are given with this act. Representations must originally be supplied through viewing, through intuitions (as we have stated repeatedly, the constructed objects of mathematics are included here):

> For through the 'I', as simple representation, nothing manifold is given; only in intuition, which is distinct from the 'I', can a manifold be given; and only through *combination* in one consciousness can it be thought.[60]

Having located the fundamental condition of judgment in the act of synthetic unity, the next step is to specify the structure of an objective theoretical judgment. To comprehend that we must draw upon another piece of conceptual apparatus which we have not yet mentioned, but which is of the utmost importance in Kant's attempt to legitimate the *a priori* synthetic judgments. This is the concept of inner sense.

Kant defines inner sense as 'the intuition of our self and our inner states.'[61] It thus has a specific psychological dimension. Kant even says that these inner senses compose our sense of self, our soul.[62] (These states are interpreted in a strictly empiricist manner. Throughout the *Critique* he criticizes those who conflate this empirical concept of self with a purely intelligible entity which ostensibly can be directly intuited.)[63] We do not view these inner states as we view other objects; we feel them. If we are to classify (judge) them objectively, then we must be able to represent them spatially, i.e. locate where they are. Even though we simply feel them, they are nevertheless necessarily governed by the condition of time, i.e. they occur in time. Hence on this basis Kant creates a division between space and time: space is the condition of outer sense; time is the condition of inner sense. But because in any judgment our mind must be affected by representations, time is also a condition of the representations of outer sense. In other words, all phenomena are governed by the condition of time, while space is limited to what can be spatially located.[64]

If a judgment is a synthesis of both types of representations, objects of intuition and logical functions, then inner sense must itself be governed by the same condition as all other representations. This all important point which Kant makes in paragraph 24 of *Critique of Pure Reason* will seem very obscure if the reader has forgotten all the other pieces of apparatus that the *Critique* has acquired. But if it is kept in mind that time is the condition of all phenomena (i.e. that time is not absolute but must be occupied by some type of intuitable object, whether empirical or mathematical, whether it be a feeling or spatially perceivable), then we can see that time itself must be dependent upon the 'transcendental act of apperception.' Or as Kant says:

> But since there lies in us a certain form of *a priori* sensible intuition, which depends on the receptivity of the faculty of representation (sensibility), the understanding, as spontaneity, is able to determine inner sense through the manifold of given representations, in accordance with the synthetic unity of apperception, and so to think synthetic unity of the apperception of the manifold of *a priori sensible intuition*...[65]

Moreover, if the pure functions of the understanding have no other objective, i.e. '*theoretically*' legitimate, use apart from the ordering of concepts which refer to phenomena, then there must be a connection between time, as the condition of all phenomena, and the pure categories, as the conditions of judgment or understanding.[66] Given that the connection is grounded in consciousness (and the 'transcendental act of apperception' has established this for Kant) it must be a faculty or capacity of consciousness. The faculty is called by Kant the 'transcendental unity of the imagination' or the faculty of figurative synthesis. It provides the answer to the question 'how are synthetic judgments *a priori* possible?' because it is the source of the synthesis of the pure forms of intuition and the pure concepts, the categories of the understanding. It makes the synthesis between the *a priori* elements derived from the 'Transcendental Aesthetic' and those derived in the first part of the 'Transcendental Analytic' possible. Those elements could only be acquired by isolating the capacity to view data and the capacity to judge it. Mathematics and the functions of logic had provided Kant with a number of clues enabling him to specify the *a priori* elements for viewing data and judging it. But only when he can find a faculty which conjoins the elements into judgments can he legitimate the *a priori* synthetic judgments. (This is why Kant does not prove mathematical judgments are *a priori* synthetic until after he has laid down the condition of synthesis and the precise foundational judgment of mathematics, 'the axioms of intuition', the afore-mentioned principle that 'all intuitions are extensive magnitudes.')[67]

What the figurative act of synthesis does is introduce the condition of time as the means of mediating between the two pieces of apparatus (sensibility and understanding) that have hitherto been examined in isolation.

Note that the faculties of pure sensibility and understanding have only been isolated in order to discover their constitutive elements. There is never any suggestion by Kant that these faculties exist in isolation.

Now if we no longer are concerned with the isolation of the *a priori* elements, but with their unity, then it will be necessary to stipulate the conditions of synthesis or co-ordination between representations. The first and overriding condition that Kant introduces is, as we have said, the formal condition of *all* phenomena — time. Hence if the categories are only functions for classifying objects of experience, if they are theoretically necessary functions for making judgment whose content is derived from observations, then they must be integrated with time. Thus the two formal conditions of phenomena and judgment taken together provide the formal condition of a judgment of a possible experience. For Kant, the 'imagination' is the source of that integration because the imagination is 'the capacity to represent an object even without it being present in intuition.'[68]

When any judgment about objects of experience is made it will be necessary to rely upon the imagination. For it will be necessary to connect a representation at one moment with the preceding moment, and such a connection is an act of consciousness. The persistent recognition of object A over the successive moments x,y,z requires a spontaneous act of consciousness at every moment. Kant calls the faculty responsible for this spontaneous act of consciousness the productive imagination. It is the capacity to affirm the persistence of an object at these different times. The same type of act of spontaneity is required even with objects in the same time, *for even the identification and specification of an object requires a selection of relations.*

Because we automatically make such connections we rarely if ever think about such acts of consciousness. Kant indicates that we are rarely conscious of our employment of this capacity, but it is an 'indispensable function of the soul'; without it we could not have knowledge. For there would be no possible judgment of experience, no conceptualization.[69]

This act of the imagination is itself subordinate to the spontaneity of the understanding, the 'transcendental act of apperception'. As we have just seen, it is also an act of consciousness enabling us to judge a specific object over successive moments of time. On the one hand it is a sub-species of the understanding. On the other hand, it is related directly to time and the faculty of inner sense. Hence, for Kant, the 'transcendental imagination' is a 'mean' between the pure categories of the understanding, and the form of inner sense.

Having found a mean between the pure categories of the understanding and the form of all phenomena, Kant has found the key to solving his problem 'how are synthetic judgments *a priori* possible?' The answer is that they are possible because of an act of synthesis which co-ordinates the forms of intuition with the functions of the understanding.

Thus far we have seen how Kant has specified *a priori* elements of objective, i.e. theoretical, judgments by locating: (1) the *a priori* elements of intuition, with mathematics as its clue; (2) the *a priori* functions of judgment, with logic as its clue; and (3) the necessity of the act of synthesis for the bringing together of any representations in the judgment. The last point has involved shifting the idea of the necessity of the judgment away from the content of specific experiences toward the conditions of the act of judgment.

The next and final methodological step in the 'Transcendental Analytic' is to specify the synthetic *a priori* judgments. The key to the step is the faculty of the 'transcendental imagination'. If time is the condition of all phenomena, and all categories must be applicable to what is in time, then any synthetic *a priori* judgment must be conditioned by time. Hence the categories must be strictly defined to conform to the conditions of time. Although time is the condition of all phenomena, it is only if something is in space that it can be objectively known. Kant points out that this is even the case with the concept of temporal succession which is objectively representable by the construction of a line. Hence although the categories of the understanding must conform to the conditions of time if they are applicable to objects of a possible experience, their objectivity requires that time itself must conform to space, i.e. the conditions of time are spatially representable. The categories which conform to the condition of time can only be objective if they also relate to objects of 'outer sense'. What they classify must have spatial existence.[70]

The specification of the synthetic judgments which are *a priori* thus involves bringing together the *a priori* elements of intuition or viewing with those of understanding. As a mean between the spontaneity of understanding and time (the condition of all phenomena), the 'transcendental imagination' is able to provide what Kant calls a schema for the categories. A schema is a rule of classification of the imagination which, according to Kant, is necessary to subsume particular representations under a universal. The schema is not identical with any particular object, but through a schema we can classify particulars. For example, there is an enormous difference between different species of dogs, and even between dogs which belong to the one species, yet we are able to classify them as dogs. This ability, says Kant, is due to the employment of a schema, a rule, not a picture — for it has no specific shape or form —, for subsuming particulars (in this case the various dogs under the universal 'dog').[71] It is the 'transcendental schema' (the capacity to subsume phenomena under *a priori* rules and judgments) that concerns Kant.

The 'categories of understanding' must conform to the specific attributes of time if they are to classify phenomena. The conformity of the pure category with the condition of time is called a schematized category. The schematized category, for Kant, provides a conceptual distinction which is very important for his project. Because the schema of the imagination restricts the pure category to phenomena, the schematized category *must* refer to objects of experience. And it is the schematization, the reformulation of a category in conformity with a temporal condition which makes a category 'theoretically' legitimate and a necessary condition of objective judgments of experience. But it must be noted that the categories of understanding prior to schematization are not so restricted. They may (and in the 'Transcendental Dialectic' Kant argues metaphysical judgments necessarily do) refer to what is not phenomenal. Yet it is initially only by virtue of their indispensability to judgments of experience that the categories of understanding have for Kant been legitimated. In other words, the real legitimacy of Kant's table of categories for Kant lies in the capacity of each category to be schematized. This is easily overlooked because of the swiftness with which Kant modifies his categories, but to miss this is to miss the whole point of the 'Transcendental Analytic.

It is only once the categories have been schematized that Kant presents 'The System of Principles of Pure Reason', the *a priori* synthetic judgments which provide the foundations for a metaphysics of science. They provide the 'canon' of judgment. They are not inductive nor empirical, but the conditions under which induction can take place. In sum, for Kant, they are what make pure natural science possible, or to use another of his formulations, they are the 'transcendental conditions' for objective judgments of experience. Rather than examine each of the specific *a priori* synthetic judgments, I shall provide a table which sets out the *a priori* synthetic principles and the major transformations of the *a priori* elements that take place throughout the 'Transcendental Analytic'. I shall then briefly discuss one of the synthetic principles of the understanding, the principle of causality, to illustrate Kant's procedure.

Table of the transformation of the elements of the 'Transcendental Analytic'

Pure categories: *a priori* elements of the understanding. When subjected to a rule of sensibility they are conditions of experience, otherwise they are empty thought forms. The corresponding logical function is in parentheses.

Time conditions imposed upon pure categories. They must be complied with, if the categories ware to apply to a manifold in intuition.

1. Quantity

Unity (Singular)
Plurality (Particular)
Totality (Universal)

1. Time-Series

2. Quality

Reality (Affirmative)
Negation (Negative)
Limitation (Infinite)

2. Time-Content

3. Relation

Of inherence and subsistence
(substantia et accidens)   (Categorical)

Of causality and dependence (Hypothetical)
(cause and effect)

Of community   (Disjunctive)
(reciprocity between the
active and the passive)

3. Time-Order

4. Modality

Possibility--Impossibility (Problematic)
Existence--Non Existence   (Assertoric)
Necessity--Contingency   (Apodeictic)

4. Time-Inclusions

Table, continued

Rule coordinating time condition with pure category	Category subsumed under rule
1. Generation (Synthesis) of time itself in the succesive apprehension of an object.	1. Magnitude. Successive addition of unity to unity, thus unity and plurality are respectively schematized. The category of totality is the complete unity in which the homogeneous units are contained.
2. Synthesis of sensation and perception with representationof time: filling of time.	2. Quality. Reality applies to a being in time --filled time. Negation applies to what is not in time, to empty time. Between being and nothing there are infinite degrees of being in time (Quantum).
3. The relationship of perceptions to each other in time. applies	3. Relation The category of substance to the persistence of the real in time. Cause and effect apply to the succession of a manifold, a sequence in time Reciprocity applies to the co-existence of determinations of one substance with another.
4. Whether and how an object is time-determined.	4. Modality Possibility applies to any representation which can appear in time. Actuality applies to what pertains to an actual time. Necessity applies to what must be in time.

Table, continued

The synthetic principles of the understanding

Mathematical principles: These apply to any pure construct of the mind as well as to phenomena.

1. *The axioms of intuition*: 'All intuitions are extensive magnitudes.'

2. *Anticipations of perception*: 'In all appearances the real, i.e. an object of sensation, has an intensive magnitude, i.e. a degree.'

   Dynamical principles. These are meaningless in theoretical judgments unless they relate to an empirical content.

3. Analogies of Experience: 'Experience is possible only through the representation of a necessary connection (*Verknupfung*) of perceptions.'

   (a) Principle of the persistence of substance: 'In all changes of appearances (*Erscheinungen*) the substance persists, and its quantum in nature is neither increased nor diminished.'

   (b) Principle of succession in time, in accordance with the law of causality: 'all alterations take place according to the law of the connection of cause and effect.'

   (c) Principle of co-existence, in accordance with the law of reciprocity or community: 'All substances in so far as they can be perceived to co-exist in space, are in thorough reciprocity.'

4. The Postulates of Empirical Thought.

   (a) What conforms to the formal conditions of experience (*Erfahrung*) is possible.

   (b) What relates to the material conditions of experience (i.e. to sensation) is real.

   (c) The real is necessary when it is determined according to universal conditons of experience.

The second 'Analogy of Experience': The law of cause and effect

Kant's legitimation of the principle 'all changes happen according to the law of the connection of cause and effect' is perhaps most clearly understood when we consider a time sequence, w,x,y, and a series of perceptions, a,b,c. It is, according to Kant, the 'transcendental imagination' which enables the production of the successive moments of time. It can only produce that succession in so far as there is an object that is viewed (either empirically or purely, i.e. a mathematical construction). In this case a,b,c refer to empirical perceptions. Now Kant makes an important distinction between two types of empirical perceptions. One is a purely subjective sequence, the other is objective. When the members of the time series and the empirical series do not stand in a necessary correlation then we are not talking about a necessary and objective sequence. Kant uses the example of perceiving a house to demonstrate a purely subjective sequence. One can observe the parts of the house in any order one wishes. In such a case the perception of a,b, or c would be equally possible at moment w. In this instance we have no objective sequence of events.

Now if we take this same house, and (to take an example that Kant does not use) we explode it, the sequence of perceptions is no longer order-indifferent. At moment w I perceive the house a, at moment x I perceive an explosion b, and at moment z there is a perception of a disintegrating house. In this case each perception stands in a necessary relationship to moments in the time series. (Note also that the time series is itself objectifiable because there is a perceptible content.) Unlike the example in which the house was merely being contemplated from top to bottom or in the reverse order, there is no way that perception c, the disintegrating house, could in this instance precede the explosion. We have here: (1) a connection between members of the time series; (2) the different perceptions; and (3) a connection supplied by the spontaneity of thought enabling us to see the relationship between perceptions a,b,c. Kant is only concerned with establishing the necessity between the time series, the empirical series, and the spontaneity of thought which connects the perceptions. He is not concerned with the empirical content of that series. A necessity exists between the time series and the order of perceptions, and the necessity is one of consciousness. For it is consciousness that binds the perceptions together. It is, for Kant, only because of consciousness that the one dimensionality of time, the infinitely divisible time-continuum is possible.[72] For Kant the principle of causality is thus *a priori* and necessary. The principle tells us nothing about specific empirical sequences. As we have said, each empirical object is a manifold of relations, and each relation is determined by the vast array of other relations in time and space. A sequence of empirical data is always a selected sequence, and there is no guarantee that this sequence will ever be repeated in nature. But for Kant there can be a guarantee that whatever is in time is necessarily dependent upon its antecedent. What exists cannot be the cause of its antecedent and every phenomenon is conditioned by its antecedent. This is consequently, for Kant, a 'transcendental condition' of experience.

The judgment is synthetic because it brings together time, phenomena and category, i.e. causality. No analysis of the word cause would of itself lead to the principle that *all phenomena* are time governed, and that all phenomena must conform to the law of cause and effect. It is *a priori* because it is not the result of induction. It is not subject to reconsideration on the basis of some event. Rather, for Kant, it is the cognitive condition of any event being objectively classified. It is not concerned with what has only logical possibility. It is a rule for a judgment of experience. It tells us nothing about the content of experience. For Kant, it enables us to explain how nature can be interpreted as conforming to law, how variables can be isolated within the time continuum, and how scientists can build up ever more complex series of empirical variables, enabling them to predict with growing precision the regularities of nature, or to delve ever further into the irregularities.

The law of causality is just one of the *a priori* synthetic principles of the understanding. But the basic procedure Kant employs to legitimate his other principles is the same. The category having been schematized is presented as contributing to a necessary rule for classifying any object of experience, but neither the category nor space and time are empirically perceptible.

Thus far we have observed that the attempt to lay legitimate metaphysical foundations has led Kant to formulate the synthetic *a priori* principles of the understanding. These principles are only theoretically legitimate in so far as they refer to phenomena, in so far as they are 'principles of the exposition of appearances.'[73] They do not refer to noumena. But if these conditions of experience are the result of consciousness, the acts of viewing and thinking, then it is possible to ask what would reality be without the cognitive conditions? All that we can say is that we don't know, because only phenomena conform to the rules of a theoretically objective judgment. To this extent Kant believes he has circumscribed the constitutive *a priori* conditions of experience. He believes he has established what rules must be conformed with if something is to be theoretically (i.e. empirically) knowable. He cannot stop people asking about the inner ground of things, what underlies the phenomena. He cannot stop people asking metaphysical questions. Moreover, in the second part of the 'Transcendental Logic', he seeks to clarify why the search for the inner ground of things, the endless striving for an absolute and final principle takes place. He seeks to explain how metaphysics is possible. Here too Kant adopts the same procedure. He isolates another cognitive faculty in order to probe what *a priori* elements accompany its very operation.

Transcendental logic part 2: The ideas of reason

The *Critique of Pure Reason* follows the path of cognition from the receptivity of sense experience, to the act of judgment and finally to the act of inference. It is this last act that provides the unity between judgments. Because of our capacity to infer we are able to reduce a variety of judgments to a smaller number of principles.[74] When knowledge can be deduced *a priori* from knowledge of the major premiss governing it, we have, in Kant's terminology, knowledge that is based upon principles. The discovery of principles which will encompass a diversity of judgments is one of the most beneficial and most complex achievements of scientific thought. Newton's law of gravitation was the most impressive example known to Kant of the discovery of an overarching principle.[75] Under it (and the laws of motion it relies upon) could be subsumed the observations made by Galileo, Kepler and Huygens of falling bodies, the orbit of Mars, the laws of centrifugal force and the swing of pendulums. Here was a principle that seemed to make all bodily interactions predictable. The search and discovery of principles is an invaluable part in the acquisition of knowledge. Once a correct principle can be formulated the advancement of knowledge in any field accelerates. The specific content of empirical principles is, for Kant, undoubtedly the result of induction. Nevertheless in the second part of the 'Transcendental Logic' he argues that just as empirical judgments were structured by a set of *a priori* rules, empirical principles are also governed by synthetic *a priori* principles.

Only an examination of the *a priori* elements and principles of reason will, for Kant, enable us to specify these other synthetic *a priori* principles. Note that if there are such principles, they will not be *constitutive* conditions of experience. Unlike a judgment which classifies sensations, an inference is a mediating activity, not between sensations but between judgments. Hence the principles that are to be sought are by their very nature not able to be schematized in the same way as the *a priori* synthetic principles or rules of the understanding. The legitimacy of a supersensible concept which cannot be schematized with the attributes of time depends upon its capacity to provide as great a systematic unity as possible for judgments of experience.[76]

There is no need for us to follow the elaborate details of the second part of Kant's 'Transcendental Logic.' More important for our purposes is to grasp how Kant arrives at the principles of pure reason, and to appreciate the scope of these *a priori* elements of pure reason.

Again Kant takes his clue from logic, but this time from the modifications he has made with the 'Transcendental Logic.' The three types of logical functions that Kant classifies as the functions of relation, provide him with the three types of syllogisms of transcendental logic: categorical, hypothetical, and disjunctive syllogisms. Given that Kant is interested not only in the capacity of reason to unify judgments, to subsume judgments under a principle, but in the ability to subsume all judgments under metaphysical principles, he is interested in locating the transition from empirical to metaphysical principles.

The types of syllogisms give Kant a clue to the type of unity and the source of the *a priori* ideas of reason. Firstly each type of syllogism exemplifies a different type of relationship between judgments. The categorical syllogism exemplifies a relationship between subject and predicate, the hypothetical syllogism exemplifies a relationship between members of a series, and a disjunctive judgment exemplifies a relationship between the different components of a system. When we move from the specific subject, series, or totality of an empirical observation, and instead ask after the unconditioned or absolute subject, the unconditioned series (or absolutely first member of the series), or the absolute totality which includes all possible parts, we move beyond the realm of experience. The search for the absolute subject, the absolute series, and the absolute totality, is, observes Kant, at the basis of the speculative sciences of rational psychology, rational cosmology, and rational theology. The most complicated details of the second part of *Critique of Pure Reason* are an attempt to demonstrate how the content of these three metaphysical disciplines is a synthesis of three unconditioned ideas (i.e. the absolute subject, series and totality) and the pure (i.e. unschematized) categories of the understanding. Kant attempts to show that each discipline involves a 'transcendental illusion', a 'dialectic' of reason. Each discipline takes the classifiers which the 'Transcendental Analytic' argued were necessary functions of experience and uses them as if they themselves revealed the attributes of the absolute subject, the absolute series, or the absolute totality. In other words, speculations about the soul, the cosmos and God are the result of the search for rational principles, and the illegitimate employment of pure concepts. In sum, the judgments of traditional metaphysics are not 'theoretically legitimate'. They cannot be verified as objectively real, hence theoretically true, because they cannot be objectively viewed. Nevertheless, claims Kant, there is heuristic value in using the ideas of an immaterial point of consciousness, of the total series of the cosmos, and of the all inclusive totality of mind, provided we do not confuse their 'immanent' employment with 'transcendent' speculations. In the former case, we remain committed to the study of phenomena, but we undertake that study *as if* there were an immaterial point of consciousness which was responsible for psychological activity, *as if* the members of the cosmos formed an infinite series, and *as if* the universe formed an intelligible and unified system of laws, i.e. *as if* the universe were the product of intelligence so that every event happened for a purpose.[77]

If, as Kant claims, there is heuristic theoretical value in the ideas of soul, cosmos, and God, then it is understandable why discourses on science could glide into metaphysical speculations without the discussants being aware that illegitimate transitions had occurred. In other words, it would be explicable how metaphysics is possible, and why metaphysics has hitherto not been a science. For no one hitherto had attempted to trace these ideas to their cognitive source. Moreover, if they are not objects of a possible experience, and their attributes exclude them being considered so, then the question of their legitimacy will require locating their source. Hitherto, rationalists have claimed they are innate, while empiricists have denied the existence of innate ideas: the only *a priori* ideas allowed by the empiricists are purely logical ones, the ideas of relations. In the 'Transcendental Analytic' Kant has attempted to clarify what gave rationalists a degree of legitimacy in their appeal to non-empirical (or in Kant's terminology synthetic *a priori*) principles. For synthetic *a priori* principles were discovered to be conditions of possible experience. But having defined the synthetic *a priori* principles, Kant circumscribed their scope in a way that defeats the very speculative aspirations which empiricism wishes to invalidate. However, no one, whether empiricist or rationalist, prior to Kant had argued that the source of the main concerns of speculative metaphysics lies in a combination of the act of making inferences and the quest for the unconditioned principle of a type of syllogism. Moreover, it could not be attempted previously because no one had undertaken a 'Transcendental Analytic'. For it is only on the basis of the 'Transcendental Analytic' that Kant can attempt to locate the precise point where a regulative maxim (a heuristic idea of research) is hypostatized and transformed into an object of speculative metaphysics whose reality and attributes can never be scientifically known. A very brief recapitulation of Kant's specification of the illicit moves in speculative metaphysics illustrates how his critique of metaphysics is dependent on the 'Transcendental Analytic'.

In the case of rational psychology, Kant argues that as soon as the act of apperception is taken as confirming the existence of an immaterial thing, an illicit move is made which leads to a transcendental 'illusion' that we have knowledge of the immateriality of the soul.

In the case of rational cosmology the speculative conundrums result from a failure to distinguish between things in themselves and phenomena, a failure that, for Kant, is a consequence of a mistaken understanding about the nature of space and time. Kant claims that in a cosmological inference the major premiss, i.e. the premiss about the unconditioned or first member of the series, is not governed by space and time, and hence not a phenomenon. However, the minor, i.e. the reference to the temporal series, does refer to phenomena.[78] Consequently rational cosmology, according to Kant, relies upon a synthesis of ontologically incommensurate concepts. In the case of rational cosmology, Kant claims one can adopt one of two opposing positions with regard to the first member of the series: one can deny it, or one can accept it. This leads to a number of antinomies which according to Kant, result from a regressive synthesis of the members of a series. The ultimate condition of a temporal series would be the first point in time; of a spatial series, the borders of space; of a composite series, the ultimate element; of a causal series, the first member of the chain; and if we inquire into the modality of the series, its first member would be either necessary or contingent. One can argue with equal consistency, claims Kant: (1) that the world does (not) begin in time and has (no) limits in space; (2) that the world is (not) composed of simple elements; (3) that there is (no) spontaneous causality, and (4) that there is (not) a necessary first cause. Whatever position one adopts, however, is metaphysical, and thus involves the same error in transcendental logic: the conflation of things in themselves and phenomena.[79]

Finally, in the case of rational theology, following Leibniz and the new physics generally, Kant holds that an object of sensation is determined by the totality of all phenomena. But the totality of phenomena is itself beyond the range of a possible experience. The totality is, in Kant's terminology, an idea. When this idea is hypostatized, and then personified by being thought of as perfect, a regulative idea becomes a transcendent one. For Kant, all speculative arguments for the existence of God are the result of a 'transcendental illusion'.

The illusion is, for Kant, grounded in a simple ontological error that the 'Transcendental Analytic' exposes. The theoretically real must be the object of a possible experience, and not merely a conceptual possibility. This ontological confusion which takes the conceptually possible and logically consistent as a sufficient condition for establishing the reality of a perfect being is, for Kant, the fundamental error of the ontological argument for God's existence. And the ontological argument is itself at the basis of the cosmological and physico-theological arguments.

Kant thus took the traditional ideas of metaphysics and denied they had any place in scientific judgment, except as regulative ideas, ideas which he traced to a cognitive source. Equally important to his undertaking was the attempt to *resite* the ideas of traditional metaphysics. Indeed it would be no overstatement to say that the stipulation of the *a priori* conditions of judgments of experience, and the demolition of speculative metaphysics are completely subordinate to Kant's attempt to resite and subsequently to legitimate on a different basis the noumenal ideas of God, freedom, and immortality. This move, which is most succinctly summed up in Kant's famous sentence 'I have therefore found it necessary to deny *knowledge*, in order to make room for *faith*', is to be explored in the next chapter.[80]

Notes

1. Note that Kant's starting point for metaphysics was with the *a priori* elements and principles of objective, or what he called theoretical, knowledge. His starting point is necessarily the analysis of the *a priori* conditions of experience, and he initially considers the supersensible ideas only if they can contribute to knowledge of experience. Only after the *a priori* conditions and scope of theoretical judgments have been circumscribed does Kant see it as legitimate to inquire into other types of judgments. For only if it were established that a type of judgment fell beyond the bounds of objective judgments of experience was it legitimate to consider the rules governing this other type of judgment. This is indicative of Kant's close alliance with the empiricists within the realm of science, an alliance that is quickly severed when he comes to consider moral judgments.
2. *What Real Progress has Metaphysics made in Germany since the Time of Leibniz and Wolff?*, p. 157.
3. In attempting to clarify the sense behind the major moves of the 'Transcendental Aesthetic' I have deviated from the order of Kant's presentation of the arguments on space and time at *K.r.V.*, B 38-40, and B 46-48. What is presented here is what I take to be the substance of the claims. I am indebted to Kuno Fischer's remarks on the fourth argument on the transcendental ideality of space at *K.r.V.*, B 47. *Op.cit.*, Vol. 1, pp. 369-370.
4. This is not only important for Kant's *Critique of Pure Reason*, but it forms the basis of Kant's *Metaphysical Foundations of Natural Science*.

5. *K.r.V.*, B 34-35. There is a technical distinction between *Wahrnehmung* and *Anschauung*. In English the general tendency is to translate the former as perception and the latter as intuition, but this translation is not universal. Caird and Körner, for example, use 'perception' to translate *Anschauung*, and their commentaries are none the worse for it. Eduard Caird, *The Critical Philosophy of Immanuel Kant*, (Glasgow: Maclehose, 1889) and Stephen Körner, *Kant*, (Harmondsworth: Penguin, 1955). The only crucial difference between the two terms is that there is no pure *Wahrnehmung*, it is always empirical (hence one of Kant's principles is called 'Antizipationen der Wahrnehmung', 'Anticipations of Perception'). Kant defines '*Wahrnehmung*' as 'empirical consciousness, that is, a consciousness in which sensation is to be found.' *K.r.V.*, B 207. *Anschauungen* can be 'pure', but they can also be empirical.
6. I. Kant, *Gesammelte Schriften*, Akademie Ausgabe, (Berlin: Reimer, 1905), Vol. 2, p. 173.
7. *K.r.V.*, B 15-16., B 205.
8. To Johann Schultz, November 25, 1788 in *Kant: Philosophical Correspondence 1759-99*, ed. and tr. Arnulf Zweig, (Chicago: Uni. of Chicago Press, 1967), pp. 128-131.
9. *The Problem Of Knowledge: Philosophy, Science and History Since Hegel*, tr. William Woglon and Charles Hendel, (New Have: Yale Uni. Press, 1950), p. 75. See *K.r.V.*, B 744. Amongst the many writings on Kant and mathematics three works should be singled out. Ernst Cassirer, 'Kant und die Moderne Mathematik "Mit Bezug auf Bertrand Russell und Louis Couturant's Werke über die Prinzipien der Mathematik"' in *Kritizismus: Eine Sammlung von Beiträgen aus der Welt des Neu-Kantianismus*, (ed.) Fr. Myrho, (Berlin: Heise, 1926), pp. 94-143. Gottfried Martin, *Arithmetic and Combinatorics: Kant and His Contemporaries*, tr. G, Plochman, (Carbondale and Southern Illinois Press, 1985). This work provides an excellent account of the debates over the analytic/synthetic nature of mathematics from Kant's time to this century, and how Kant's account of mathematics has been (largely mis)interpreted. A work looking at more contemporary accounts of mathematics in light of Kant's writings is Klauss Mainzer's brilliant Ph.d. *Kants Philosophie: Begründung des mathematischen Konstruktivismus und seine Wirkung in der Grundlagenforschung*, Inaugural Dissertation der Philosophischen Fakultät der Westfälischen Wilhelm-Universität zu Münster, WS 1972/1973.
10. This illicit move from mathematical reasoning to metaphysics is a recurring theme in Kant. Plato is held by Kant as the originator of this move. Plato is, for Kant, the father of enthusiasm (*Schwärmerei*). Leibniz and Newton

are also guilty of 'free flight' once they have 'risen' with mathematics above experience. The low regard that Kant has for spurious rationalist metaphysics, most evident in the pejorative term *Schwärmerei*, is offset by the high regard for what he considers to be the lasting greatness of rationalism, its contribution to moral thought. See esp. *K.r.V.*, B 8-9; B 370-375; *K.d.U.*, Pt. 2 para. 1. For an interesting interpretation of Kant which makes the problem of curing us of 'enthusiasm' Schwärmerei compared to Kant's project see Robert Butt's *Kant and the Double Government Methodology: Supersensibility and Method in Kant's Philosophy of Science*, (Dordrecht: Reidel, 1984).
11. Speaking of the transcendental principle that 'all intuitions are extended magnitudes', Kant claims: 'it alone can make mathematics, in its complete precision, applicable to objects of experience.' *K.r.V.*, B 206.
12. *Prolegomena*, para. 10.
13. *K.r.V.*, B 49-52.
14. Hence Kant's concern to emphasize that he is no traditional idealist when he is making his case for the transcendental ideality of time and space. *K.r.V.*, B 68-70. Any psychological description must be empirically verifiable, and this can only be the case if the object of its inquiry is in time. Time is a transcendental condition, not a psychological one. By making this distinction Kant is erecting barriers against the rational psychologists, whom he sees as moving illicitly from a transcendental and legitimate employment of the concept of the subject to a speculative one, which is not under the condition of time or space, and therefore not an object of scientific inquiry. However, the cognitive framework which Kant employs was not without problems of its own, as we shall discuss later.
15. This is particularly crucial for the 'Analogies of Experience.'
16. See *K.r.V.*, B 156, B 292.
17. *K.r.V.*, B 744.
18. As this example and other remarks in the *Critique* illustrate, the properties of space Kant has in mind are obviously Euclidean. Now Kant knows that the construction of the concepts depends upon definitions, definitions which he claims are precise in a way that other concepts are not. Mathematical concepts, unlike empirical concepts, are 'first given through the definition' (*K.r.V.*, B 759). But he sees the expansion of those definitions as governed by the form of space, and that form is presented as the condition of construction, rather than a moment *within* construction. Hence its properties are not only necessary but determinate of the possible constructions. This is what makes it so difficult to reconcile Kant's theory of space with the general theory of relativity in which space and space-time are

curved. Non-Euclidean geometry is used to measure gravitational effects not adequately accounted for by Newtonian physics. There has been much written over whether Kant's conception of space can accommodate this turn of events, e.g. J. E. Wiredu, 'Kant's Synthetic A priori in Geometry and the Rise of Non-Euclidean Geometry' in *Kant Studien*, Heft 1, 1970; D. P. Dwyer, *Kant's Solution for Verification in Metaphysics*, (London: George Allen and Unwin), pp. 160-169. See also Wilhelm Meinecke, 'Die Bedeutung der nicht-Euklidischen Geometrie in ihrem Verhältniss zu Kants Theorie der mathematischen Erkenntnis', *Kant Studien*, 1906, XI. But Kant simply does not address the problem of non-Euclidean geometries being employed in physics. His conception of space is not only conceived directly in response to problems of classical mechanics, as we have stressed by focussing upon the principle of the continuum and the infinite divisibility of matter, but it is tied up with problems of metaphysics that would now be considered to have no necessary connection with physics. In other words Kant imposes a different type of agenda from that of the working physicist or mathematician. Because of this Kant's philosophy is defenceless against new advances in physics which drastically transform the framework as well as the content of physics.

19. See *On the Form and Principles of the Sensible and Intelligible World* (Inaugral Dissertation 1770) in *Kant: Selected Pre-Critical Writings and Correspondence with Beck*, tr. G.B. Kerferd and D.E. Walford, (Manchester: University Press, 1968), para. 15 D.

20. *Metaphysical Foundations of Natural Science*, tr. James Ellington (Indianapolis: Bobbs-Merrill, 1970), p. 20. On p. 125 space without matter is said to be 'a necessary concept of reason, and is nothing but a mere Idea.' There is nothing derogatory for Kant in something being 'a mere Idea'. Kant is concerned to eliminate incorrect ontological classifications. For an excellent account of how Kant's conception of relative space functions for observing matter in motion see Robert Palter, 'Kant's Formulation of the Laws of Motion' in *Space,Time and Geometry*, ed. Patrick Suppes, (Dordrecht: D. Reidel, 1973).

21. Leibniz's aversion to Newton's absolute space and time, is fully in keeping with his insistence that only concrete individuals exist, which is the underlying idea of the doctrine of monads. Leibniz's nominalism is well brought out in chapter X of Benson Mates's *The Philosophy of Leibniz: Metaphysics and Language*, (Oxford: University Press, 1986). See also chapter XIII on 'Space and Time.' The difference between Leibniz and Newton over the nature

of space and time is at the centre of the controversy between Leibniz and Clarke.
22. *On the Form and Principles of the Sensible and Intelligible Worlds*, sect. 15 D. See also para. 14 sect. 5 for a critique of Leibniz's conception of time.
23. *Prolegomena*, para. 13.
24. In his critique of Eberhard, Kant calls the *Critique of Pure Reason*, 'the genuine apology for Leibniz, even against his partisans whose eulogies scarcely do him any honour.' The *Kant-Eberhard Controversy*, tr. Henry E. Allison, (Baltimore: John Hopkins Uni. Press, 1973), p. 160.
25. See esp. *Metaphysical Foundations of Natural Science*, pp. 54-56. Here Kant suggests that Leibniz's doctrine had been badly misunderstood, that his doctrine of monads was not meant to be an explanation of 'natural appearances' (*Naturerscheinungen*). It was a Platonic concept, referring to things in themselves.
26. This is the problem of the 'second Antinomy' of the 'cosmological ideas.'
27. *K.r.V.*, B 34. For Kant the term 'appearances' has nothing to do with something not being real. Indeed the first half of *Critique of Pure Reason* seeks to establish that if something objectively possesses the predicate of existence or reality, it must appear (*erscheinen*). Even apart from this all important point, Kant goes to great length to distinguish this term from any thing to do with a sensory illusion or solipsism. Note his exasperation with the Garve/Feder review of the *Critique*. See the 'Appendix' and notes two and three to para. 13 of *Prolegomena*, and *K.r.V.*, B 69-70.
28. For Newton see the General Scholium to *Principia* and Queries 28 and 31 of the *Opticks*, in *The Leibniz-Clarke Correspondence*. The distinction is at the basis of Locke's division between nominal and real essences. See esp. *Essay*, Vol. 1. pp. 402-403.
29. *New Essays*, p. 397.
30. E.g. *New Essays*, p. 120, p. 403.
31. This connection is constantly made by Leibniz. In the *Monadology* we can trace step by step how Leibniz arrives at the moral world. The connection between the moral reasoning of the *Theodicy* and questions of physics also plays an important part in Leibniz's 'Fifth Paper' in the correspondence with Clarke.
32. The overwhelming importance of Leibniz for Kant's enterprise leads Kant to devote a chapter to a critique of Leibniz as an appendix to the first half of the *Critique*. That chapter attempts to demonstrate how Leibniz's metaphysical errors all derived from his failure to have located the transcendental source and hence the legitimate scope of his 'Concepts of Reflection.' Gottfried Martin is not

overstating the case when he says that Kant's procedure 'takes the form of a continuous discussion with Leibniz.' *Kant's Metaphysics and Theory of Science*, tr. P. G. Lucas (Manchester: University press, 1955 [1951]), p. 1. Martin's book is an excellent attempt to mediate between Cohen's and Heidegger's insights on Kant.

33. *Critique of Practical Reason* (Hereinafter cited as *K.p.V.*), tr. Lewis White Beck, (Indianapolis: Bobbs-Merrill,1956), p. 97. Page numbers cited are those of the Akademie edition which Beck inserts in brackets in the text and as running heads. For a reader unfamiliar with the context of Kant's undertaking it is bewildering why in the 'Transcendental Aesthetic', in the midst of a discussion about time, Kant launches into a critique of Leibniz's and Wolff's distinction between moral (intelligible) ideas and appearances. *K.r.V.*, B 60-63. The reason is that the problem of moral freedom is at the centre of the architectonic of the critical philosophy. The dispute with empiricism and rationalism extends across the boundaries of physics, metaphysics, and ethics, and as he later realized, aesthetics. This should not be surprising if, as Kant claims, all four disciplines draw upon the *a priori* elements and principles of pure reason.

34. In *K.p.V.*, p. 53, Hume is said to have taken the objects of experience for things in themselves 'as is almost always done.'

35. This difference is fundamental to the entire 'Transcendental Dialectic.' But the difference is discussed specifically at B 493-499.

36. Note that for Kant there is only as much science in our understanding of nature as there is mathematics. *Metaphysical Foundations of Natural Science*, p. 6. See also p. 9. And mathematics is dependent upon synthetic judgments which are *a priori*.

37. *K.r.V.*, B 451-452, B 513-514.

38. The fact/value distinction is often traced back to Hume. Hume does make this distinction, nevertheless he bases moral judgments on judgments of facts, i.e. facts concerning human nature and common sentiments and inclinations. *Enquiries*, pp. 221-3. As with Leibniz the judgments differ by degree. It is interesting to note that many logical positivists not only found an ally in Hume's desire to eliminate metaphysics, but they shared his conception of morality, which at once refers moral theories to matters of facts while at the same time making morality dependent upon custom, habit and the ensuing personal inclinations. It is difficult to see how Hume and empiricists like him can avoid moral relativism. And this is at the root of Kant's anti-empiricist moral theory. For Hume as an ancestor of twentieth century logical positivism see Rudolph Carnap,

'On the Character of Philosophical Problems' in *Philosophy of Science*, Vol. 51, No. 1, March 1984, [1934]. One of the best known works on ethics in the logical positivist tradition is Moritz Schlick's *Problems of Ethics*. According to Schlick, 'Kant's categorical imperative, which demands that one act wholly independently of one's inclinations, demands what is impossible... Moral conduct is either impossible or it is derived from natural inclinations.' tr. David Rynin, (Dover: New York, 1962), p. 62. Lewis White Beck has correctly pointed out that there is also important overlap between twentieth century positivism and Kant. 'The points of resemblance are: no categorematic concept or synthetic judgment is legitimate if it cannot be exhibited, directly or indirectly, in experience, and any judgment which cannot be verified in experience can be exposed as a misclassified analytic judgment or as erroneous empirical judgment.' *Early German Philosophy: Kant and his Predecessors*, (Cambridge Mass.: Harvard Uni. Press, 1969), p. 483. See also Beck's 'Translators Introduction' to *Critique of Practical Reason*, (New York: Bobbs-Merrill, 1956), p. XV. But, from Kant's perspective, positivism leaves out a dimension of being human which Kant held to be unrefuted by empirical science and too important to be omitted.
39. *K.p.V.*, p. 71.
40. *K.r.V.*, B 52.
41. This divine intelligence is called by Kant an *intellectus archetypus*. Its importance, for Kant, lies in its heuristic value as a comparative concept with our intelligence, an *intellectus ectypus*. The contrast between these two types of intelligence extends to every corner of the critical philosophy. See esp. *K.d.U.*, para. 77.
42. See esp. *K.r.V.*, B 93-94.
43. *K.r.V.*, B 94, 106. Later he will introduce a distinction between determinant and reflecting judgment and attempt further to demarcate the *a priori* scope of the latter.
44. *K.r.V.*, B 87.
45. *K.r.V.*, B 84-85.
46. *K.r.V.*, B XXIV-XXV.
47. *K.r.V.*, B 90-91.
48. On the basis of a survey of logic texts between 1700-1770 Erich Adickes demonstrated that Kant's list of functions had no predecessor. See *Kants Systematik als systembildender Factor*, (Berlin: Mayer and Müller, 1887), p. 33 ff. This should not be surprising. For as Adickes points out: 'Again, Kant does not simply focus on the form of the judgment, rather he distinguishes them on the basis of their content.' My translation, p. 36. I have reproduced the list of logical functions in the table on p.87, where I present the major

transformations of the *a priori* elements involved in Kant's specification of the *a priori* synthetic judgments.

49. In paragraph 9 of *K.r.V.* Kant draws attention to the following modifications. In formal logic a singular judgment can be treated as a universal judgment, but in the acquisition of knowledge it is essential to draw a distinction between the two. Likewise, he claims that an affirmative judgment in a transcendental logic must be distinguished from an infinite judgment, even though in general logic this would not be necessary. For in the acquisition of knowledge it is useful to take account of the fact that a negation leaves open an infinite number of other possible predicates which may be drawn upon for consideration in place of the negated predicate. Hence the 'negative' facilitates deriving a predicate from another 'sphere' of possible predicates. At the same time the sphere is limited by its negation, i.e. the predicates which cannot be ascribed to the subject. Also fundamental to a transcendental logic and a complete table of logical functions is the distinction between hypothetical, disjunctive and categorical judgments.

50. The 'categories' are also set out on p.87. Kant believed, wrongly, that his readers would automatically see the connection between the list of logical functions he provides and the table of categories. He saw that the reader's perplexity may have been the relationship between disjunctive judgments and the category of community or reciprocity. In a disjunctive judgment we have the concept of a sphere containing a number of judgments which are not subsumed under each other. They are co-ordinated with each other 'so as determining each other, not in one direction only, as in a series, but reciprocally, as in an aggregate — if one member of the division is posited, all the rest are excluded, and conversely.' (*K.r.V.*, B 112.) This he claims is what we have in the concept of reciprocity. It is not to be mistaken for a causal relation, which Kant correlates with hypothetical judgments, for there is no subordination between members in a series. Rather a totality of substances is thought in which each exists independently of the whole, yet their total interaction is combined in the whole.

51. A 'transcendental deduction' is the legitimation of an *a priori* concept. That legitimation involves locating the precise source of a concept. Kant points out that he takes the concept of 'deduction' from legal parlance which distinguishes between the fact that an event takes place and the right or legitimacy of an action. This distinction between the *quid juris* and the *quid facti* permeates the entire *Critique*. As we have seen the *Critique* starts with the *fact* of mathematical science, and then seeks to discover

its transcendental or *a priori* and necessary conditions. See *K.r.V.*, B 116-117, 120. Note also that Kant has not provided a transcendental deduction of the categories when he initially lists them. Paragraph 14 bears the title 'Transition to the Transcendental Deduction of the Categories.' The initial presentation of the logical functions is a 'metaphysical' not a 'transcendental deduction'. Hence it is only a clue or a guide (*Leitfaden*) to a solution. If this is not recognized all sorts of complications are created which have little to do with Kant's undertaking. One of the most interesting complications has been the search to legitimate Kant's claim that the logical functions he presents are complete. Klauss Reich's *Die Vollständigkeit der kantischen Urteilstafel*, (Berlin: Richard Schötz, 1948) is a remarkable attempt to demonstrate the logical affinities between the types of functions listed by Kant. As Reich demonstrated, the problem was that such a necessity was essentially a logical not a 'transcendental deduction'. The result was that Reich's solution had more in common in its approach with Fichte and Hegel than with Kant. Although the logical functions provide the '*logical*' starting point of Kant's presentation, it is far more likely that Kant had started with the *a priori* synthetic judgments, and the *a priori* elements and then set to work on modifying the logical functions. A number of scholars have pointed this out, e.g. Oscar Ewald, *Kants kritischer Idealismus*, (Berlin: Ernst Hoffman, 1908), pp. 28-29.

52. The following passage is crucial to Kant's strategy: 'The knowledge yielded by understanding, or at least by the human understanding, must therefore be by means of concepts, and so is not intuitive, but discursive. Whereas all intuitions, as sensible, rest on affections, concepts rest on functions. By "function" I mean the unity of the act of bringing various representations under one common representation. Concepts are based on the spontaneity of thought, sensible intuitions on the receptivity of impressions.' *K.r.V.*, B 93.
53. *K.r.V.*, B 130.
54. *K.r.V.*, B 130.
55. *K.r.V.*, B 134.
56. *K.r.V.*, B 130.
57. *K.r.V.*, para. 16.
58. *K.r.V.*, B 399, B 404.
59. The conflation of the act of apperception with a real 'thing' is *the* mistake which Kant believes is at the heart of Cartesian idealism, and which he makes responsible for the 'paralogisms' committed in rational psychology.
60. *K.r.V.*, B 135. It should be apparent that it is inappropriate to transform Kant's conception of the 'transcendental

subject of apperception' into a psychological, sociological or anthropological construct. But this conflation is not uncommon. For example, the French structuralist Marxist, Louis Althusser does this when he praises Marx's theory of *praxis* as breaking beyond the concept of the human essence. According to Althusser, philosophy prior to Marx was incapable of moving beyond the idea that there is an essence in each individual subject. Marx, according to Althusser, was able to replace the opposition between the 'empiricism of the subject' and the 'idealism of the essence' with the concepts of the forces and relations of production etc. See *For Marx*, (London: Verso, 1977), tr. Ben Brewster, pp. 227-228. This is a variant of the socio-historicizing of knowledge which dissolves the problem of the rules and forms, the 'logic' of scientific thought, into the historical conditions under which that thought occurs. Such a procedure not only does not answer Kant's question, invariably it assumes Hume's primacy of custom, expanding the concept to include socio-economic determinants. Simultaneously it moves the focus of the discourse away from the structure of judgment and toward socio-economic mechanisms. A similar type of reduction is found in Nietzsche's critique of the subject and his critique of Kant. Nietzsche believed that the ego is only ever an empirical (a mixture of a physio/psycho/sociological) perspective. Within this context Kant's problem is dissolved with the assertion that what is true or false depends upon what is life expanding. In other words a problem of the structure of judgment is replaced by a biological one. See *Beyond Good and Evil,* para. 11, cf. para. 4, 6, 12, 16,17. The position which informs Nietzsche's critique of Kant is, interestingly enough, based upon a rejection of the distinction between the thing in itself and appearances. Kant may have been fascinated to find the motifs of Nietzsche's thinking — strict anti-rationalism (devotion to appearances), scepticism (nihilism), voluntarism (will to power), and moral relativism (master/slave morality) — all coinciding with a lack of analysis of the structure of types of judgments, and a ridiculing of the question 'are synthetic judgments *a priori* possible?' There is some attempt on Nietzsche's part to provide empirical explanations for the categories of identity and equality. But these are not systematically developed, nor does he come any closer than Locke did to explaining how an axiomatic science can be applicable to the contingent world. See *The Gay Science*, para. 111, 112; *Human all too Human,* pt. 1, para. 16-20; and *On Truth and Lie in the Extra Moral Sense.*
61. K.r.V., B 49.

62. *K.r.V.*, B 51. In paragraph 24 of the *Anthropology From a Pragmatic Point of View* Kant defines inner sense as follows: 'Inner Sense is not pure apperception, consciousness of what we are *doing*; for this belongs to the power of thinking. It is, rather, consciousness of what we undergo insofar as we are affected by the play of our own thoughts.' tr. Mary J. Gregor, (The Hague: Martinus Nijhoff, 1974).
63. Note that Kant takes great pains to distinguish the transcendental ego, the 'I think', from the empirical self. Aware of the possible confusion he draws attention to this distinction in the midst of the 'transcendental deduction'. *K.r.V.*, B 153-156.
64. *K.r.V,* B 50-51.
65. *K.r.V.*, B 150.
66. *K.r.V.*, para. 22 states the all important condition 'the categories, as yielding knowledge of things, have no kind of application, save only in regard to things which may be objects of possible experience.' It is here Kant also makes the claim that is so important for his entire conception of mathematics: mathematical concepts do not of themselves qualify as knowledge. They only qualify in so far as they stand under the forms of intuition. *K.r.V.*, B 147. In other words, their applicability is part of their very nature, and this is what distinguishes mathematical concepts from mere poetic figments. Mathematics, as we have indicated before, is not a means for speculative ascension to a metaphysical realm. Rather, as Descartes had stated, it is a means for becoming 'master over nature.' *K.r.V.*, B 753.
67. Ernst Cassirer points this out against Louis Couturant and Bertrand Russell in his (qualified) defense of Kant's conception of mathematical judgments as synthetic. 'Kant und die Moderne Mathematik (Mit Bezug auf Bertrand Russell und Louis Couturant's Werke über die Prinzipien der Mathematik', *op.cit.*, p. 132. The essay is particularly interesting in light of the fact that it appeared at a time when the attempt to explain mathematics as a *purely* logical science was vigorously being pursued.
68. *K.r.V.*, B 151. The importance of the imagination as a cognitive faculty coincides with the employment of scientific models (hence, also its importance in Descartes). That the imagination holds the key to the solution for the possibility of theoretical judgments would have seemed extraordinary to the ancients, just as it seems extraordinary to us, if we fail to discern the terrain upon which Kant's problem and solution arise. The way Kant poses and answers his problem is so closely allied with classical mechanics and scientific models that it simply fell beyond the scope of Plato or Aristotle.
69. *K.r.V.*, B 103.

70. *K.r.V.*, B 154, 276, 292.
71. Indeed, as Kant indicates, it is by means of a schema that pictures may be employed in a geometrical demonstration. With geometry the picture is subordinate to the schema and not the other way around. *K.r.V.*, B 177-181.
72. See esp. *K.r.V.*, B 243-244. Of course, once the foundations of classical mechanics, the classical concepts of space and time, the continuum and the concept of the mass point are revised, Kant's 'Transcendental Deduction' and the 'Transcendental Aesthetic' no longer have any claim to providing the conditions of experience. The fate of Kant's philosophy is inseparable from the fate of the science whose foundations it sought to legitimate. The application of non-Euclidean geometry in relativity theory as well as the intrinsic connection between velocity and space-time had made irrelevant the conception of space underpining the 'Transcendental Aesthetic.' The implications of quantum physics for the continuum created an insurmountable problem for the foundations of Kant's metaphysics of experience. Perhaps it could be shown that classical physics supplies the epistemic foundations for post-classical physics. But such an attempt would do nothing to advance physics. See ch. 9 of Ernst Nagel's *The Structure of Science: Problems in the Logic of Scientific Explanation*, (London: Routledge and Kegan Paul, 1961) which, amongst other things, points out the dangers of trying to make all geometries derivative from Euclidean geometry. Even aside from unforseen developments, the architectonic of Kant's philosophy and his employment of cognitive faculties meant that his influence on theoretical physicists owed little to his transcendental idealism. This is evident, for example, even in the case of Hermann von Helmholtz, a major physicist who found some inspiration in Kant. See his *Epistemological Writings*, ed Paul Hertz and Moritz Schlick, (Dordrecht: Reidel,1977). The significance of twentieth century physics for Kant's philosophy is well brought out in Ernst Cassirer's *Determinism and Indeterminism in Modern Physics*, tr. O.T. Benfey, (New Haven: Yale Uni. Press, 1956). There is also an interesting discussion on Kant's concept of causality and quantum mechanics reported by Werner Heisenberg. 'Quantum Mechanics and Kantian Philosophy' in *Physics and Beyond: Encounters and Conversations*, (London: George Allen and Unwin, 1971). It would be an oversimplification to argue that these developments have proved Hume right at Kant's expense. For classical physics, Kant is able to account for far more than Hume.
73. *K.r.V.*, B 303.
74. *K.r.V.*, B 355-359.

75. The point is well made by Ernst Cassirer in *Kant's Life and Thought*, tr. James Haden, (New Haven: Yale Uni. Press, 1981), pp. 291-2.
76. *K.r.V.*, B 693, 697-698.
77. *K.r.V.*, B 710-14.
78. *K.r.V.*, B 527-528.
79. For Kant, these 'antinomies' of pure reason exemplify the metaphysics of rationalism and dogmatism. The thesis is essentially rationalist, the antithesis empiricist. Kant's resolution of the antinomies involves the claim that in the first two antinomies both arguments are false. The conception of infinite divisibility as a mathematical concept dependent upon the transcendental conditions of space and time plays the key to the their resolution. In the last two antinomies, Kant holds that we can think both positions as true.
80. *K.r.V.*, B XXX.

3 The transcendental foundations of absolute freedom and the site of faith

To appreciate why Kant sought to locate a moral site beyond the realm of knowledge, it is necessary to consider the enormous influence that Rousseau played on Kant's thinking about morals. For while the Newtonian world picture had supplied the clues for Kant's attempt to lay a metaphysics of experience, it was Rousseau who drew Kant's attention to the importance of defining the boundary of experience in order to save the idea of absolute moral freedom.[1]

Rousseau: The need for a transcendent will

If Descartes had sought to set humanity on a better course by seeking salvation in the progress of science, it was Rousseau who most passionately questioned the value of that path. He saw that there was not only no necessary connection between liberty and the technical mastery of nature, but progress in science and the arts had a corrupting influence upon the moral sensibilities of the moderns. This was obvious, for Rousseau, in the fact that civilization was not only consolidating the rule of the rich and powerful, but it was creating inequalities and injustices on a scale hitherto unknown.

For Rousseau, then, the problem of modernity was not the Cartesian one of the satiation of human needs by the instrumentalization of the world, rather it was the moral one of reawakening moral forces that civilization had corrupted. That reawakening required the employment of a capacity which the mechanistic philosophy, with its emphasis upon 'facts', had eliminated — the immaterial concept of the free will. In its place mechanistic philosophy had concentrated upon the mechanistic determinants of the will: the instincts, the passions, and the environment. Rousseau believed that in so far as the moderns live in a corrupted environment and have corrupted passions, contemporary philosophers necessarily observe corrupted behaviour. On the basis of their observations they produce a picture of human beings as appetitive, acquisitive and egoistical machines. Their picture of human beings makes chimeras of the ideas of justice and virtue. Within this model, public and private virtues are pretexts for force.[3] Hence, claims Rousseau, those who base their political solutions on facts arising within corrupted civil society may easily aid the cause of tyranny.[4]

For Rousseau the idea of the free moral will had to be salvaged because the only way to create a just society is to use an uncorrupted foundation. The free moral will, for Rousseau, could serve this purpose, because the free will is by definition necessarily unconditioned. The formulation of duties by the will is not mechanical but an act of judgment, a 'spiritual' act. It is thus beyond the bounds of mechanism.

> For physics may explain, in some measure, the mechanism of the senses and the formation of ideas; but in the power of willing or rather of choosing, and in the feeling of this power, nothing is to be found but acts which are purely spiritual and wholly inexplicable by the laws of mechanism.[5]

In opposition to the determinists who interpret the will as an epiphenomenon of the passions, Rousseau finds in the will the possibility of the subordination of passion. Not only is the will the real source of virtue, but virtue consists in obedience to the prescription of the will and not one's inclinations: 'there is no virtue in following your inclinations and indulging your taste for doing good just when you feel like it; virtue consists in subordinating your inclinations to the call of duty.'[6] Rousseau believed that the conception of virtue he was putting forward was the basis of common moral practice. The mechanistic philosophers had overlooked this common moral practice by trying to explain it in on the basis of their mechanistic picture of human beings driven by self-interest.[7]

With the conception of the transcendent will, a will common to all, Rousseau believed he had located the source for founding universal moral principles, and thus for creating a social order based on virtue. Its underlying principle, the general will, was that every person be autonomous in formulating what is good, but the good must be formulated in such a way that it is universal and binding on all. Each individual is legislator and subject. Freedom, thus conceived, consists in human beings complying with a law that they themselves prescribe, and a just society will be one whose institutions reflect the free and general will.[8]

It should be noted that the cleavage between the objects of mechanism and the exercise of the free will in no way involves a rejection of the results of the empirical studies of the passions, inclinations or environment. On the contrary, Rousseau not only concedes that human beings are shaped by instincts, passions, and the environment, he himself is concerned to reshape those instincts as well as the environment in which human beings live. But the reshaping of behaviour is to be governed by the idea of freedom of the will. The ideas of good and evil are to be grounded in the will, not in specific social practices which themselves stand in need of evaluation. The new science may well reveal the mechanisms of nature, but it does not reveal the end or destiny of human beings because it has no coherent way of dealing with concepts such as unconditional duties or rights.[9] For this reason the boundaries of science need to be defined and the conception of moral freedom needs to be safeguarded.

In all of these points, and many more, Kant was the disciple of Rousseau.[10] In morals Kant begins with the goodness of the will, and he derives his concept of duty (*Pflicht*) from the will. Like Rousseau he makes justice depend upon the concept of moral autonomy, and justice becomes the yardstick for measuring the worth of social practices. Like Rousseau he opposes Hobbes's subsumption of natural law under positive law (though Kant's conception of political sovereignty is far closer to Hobbes than Rousseau).[11] Like Rousseau he vigorously attacks any empiricist morality as destructive to the universality of moral law (though Kant's cosmopolitan political vision departs considerably from Rousseau and is much closer in spirit to Diderot). Like Rousseau, he espouses a moral theism as he equates moral duties with divine commands. Like Rousseau he separates the autonomous will from the inclinations and passions, making autonomy rather than heteronomy the ground of freedom. Thus like Rousseau he creates a cleavage between the mechanistic world of science (the 'realm of understanding') and the autonomous sphere of the will (the 'realm of practical reason').

Once Kant had specified the source and types of the primary *a priori* conditions of experience (*Erfahrung*) or necessity he could know the limits, the scope, of those principles. Having believed he had established those limits, his next step was to specify the transcendental conditions of a judgment of freedom. This necessarily involved specifying: (a) the *a priori* conditions of such a judgment (its elements and hence its source); (b) that such conditions were different in kind from those of judgments of experience (its scope); and (c) the status of a judgment purporting to be a judgment of freedom.

Kant's legitimation of moral freedom

A major theme of the first half of *Critique of Pure Reason* is that we do not and cannot view or intuit intelligence. We view phenomena. Hence we cannot have scientific, theoretical knowledge of intelligence as something distinct from phenomena. At most we may use the concept of pure intelligence as a regulative principle.[12] Hence, for Kant, we can never have empirical proof of the causality of intelligence, for we could only *observe* the phenomenal effects of intelligence.

Although science is not able to prove the legitimacy of an intelligible causal principle, moral experience, observes Kant, relies upon the idea that human beings do judge themselves and others on the basis not only of what they do, but on the principles guiding their actions. For example, if someone appears to be generous, but it is later discovered that the actions of generosity were done for the sake of private gain, the worth of the action is judged differently, even if some people have genuinely benefited from the act.

If people do seek and do refer to the principles governing behaviour, it is no use explaining this type of judgment away. It may be that in showing interest in the principle of an action rather than in the environmental and physiological conditions which determine a human being, people are getting caught up in an illusion. It may be the case that people are automata driven by the instincts and social messages at any given moment.[13] But to assess the worth of the principle governing the action, people believe that a principle may govern an action. It is often the principle and not just the act which people want to consider. We may also wish to consider the circumstances determining the action and, for Kant, these alone provide the objective content of a theoretical judgment, and thus of scientific inquiry.[14] But people are capable of separating the principle of the will from the particular circumstances governing a particular action. Thus, for example, if we believe someone genuinely attempts yet fails in the performance of a moral task, we still esteem the person on account of the moral intention. In making this separation, Kant observes that we believe our principles may supply the grounds for actions. For Kant this means we believe in the possibility of intelligence as causal. And we believe ourselves to be not only subject to law, but legislators; we believe ourselves to be intelligences as well as composites of forces and instincts. In Kant's terminology, by believing this we think ourselves members of an 'intelligible realm'. By so doing we have moved beyond the scope of theoretical reason and science. The moral empiricists see this move as illegitimate, but, for Kant, they are failing to grasp an important dimension of being human: the capacity people have to make ideals and to measure themselves by them.

It is in the formulation and evaluation of principles of the will that a different type of judgment, a practical or moral judgment, is set up alongside a theoretical judgment: the one is concerned with principles of the will, the other with actions; the one is prescriptive, the other descriptive; the one does not restrict itself to what we are, but asks what we *should* be. The will as the source of maxims and principles is also, for Kant, the necessary source of the judgment. The problem is to see whether the will is able to formulate principles that we believe should command compliance from everyone who participates in the 'realm of intelligence', i.e. from everyone who inquires after and wishes to communicate the intelligible ground, the rational intentions, of their actions. Such compliance is not that required in a theoretical judgment. It is not a matter of demonstrating a fact, but rationalising and evaluating intentions. The legitimacy of such a judgment is based on concepts and 'mere reasons.' The actions must be viewed in light of the reasons, and not the other way around. Moreover the ground of compliance must, for Kant, itself be non-sensuous if it is to be universal and not merely particular in its scope.[15] Thus from the outset Kant is involved in a critique of all moral theories which make anything other than reason alone in its practical employment the ground of moral obligation. Particularly pertinent to Kant's plan of attack is his critique of *eudaimonian* ethics.

Kant's critique of happiness as the supreme principle of moral obligation is essentially that happiness is by its very definition particular, and contingent. Not only are different people made happy by different things, but our own desires are constantly changing, so it would, says Kant, be impossible to formulate a universal law of happiness.[16] Along similar lines, the sacrifice that individuals make in fulfilling moral obligations, may bear little or no relation to personal happiness. Our esteem is generally proportionate to the personal sacrifice which is required to fulfil one's moral obligation.[17]

From the social perspective happiness is also inadequate as a ground of practical reason. Rousseau had pointed this out in his *Discourse on the Origins of Inequality*. He had seen that in a commercial society 'we find our advantage in the misfortunes of our fellow-creatures and...the loss of one man almost always constitutes the prosperity of another.'[18] In such a state it is hard to understand why happiness should be the basis of political morality. For it is almost impossible to imagine a political decision which will make everyone happier than before. Certainly, this is a most unlikely outcome in political decisions which require substantial redistributions of limited resources.

Kant, like Rousseau, had no reason to believe that the members of the aristocracy whose privileges would be removed under a just (i.e. a republican) constitution would be happier. Even if one accepted the utopian fantasy that scarcity of all resources and offices could be eliminated, there is no way of knowing that all members of society would be happier. The satiation of physical desire is no guarantee of psychological satisfaction. In addition to the problems that scarcity and the elusive nature of happiness create for happiness as a universal moral guide-line for political decisions, there is the further problem that people may even be happier under a despotic government.[19]

For these reasons Kant concurs with Rousseau, that justice is the most important public good and within the private and public spheres morality and justice must take precedence over happiness.[20] This does not mean happiness is not important for Kant. On the contrary, he claims it is natural to want happiness. Thus the highest good for us *must* include happiness, but happiness must be subordinate to the concept of morality.[21] As Kant says, 'morals is not really the doctrine of how to make ourselves happy but of how we are to be *worthy* of happiness.'[22]

In discussing Kant's moral foundations it is important not to underestimate the pervasive presence that the Christian existential vision of the human condition plays in his thinking. This vision is most conspicuous in his idea that the highest good leads us to postulate the existence of the two pillars of Christianity — a highest being who dispenses justice on the basis of merit and the immortality of the soul. Kant demands that in this life we must be prepared to modify our happiness to conform with what is moral and just. Only in an afterlife is complete reconciliation possible. And there can be no knowledge of an afterlife, only faith. That faith, like all Christian accounts of faith, is only meaningful within the context of an unconditional will. But Kant's transference of faith onto a rational plane and his grounding of the will in a faculty of pure reason, a reason comprehensible to the human mind is indicative of a radical departure from the Judaic and Pauline emphasis upon the incomprehensibility of divine command. The moral theism that coincides with Kant's representation of the human condition and moral salvation is but the elevation of reason itself, the reason which, for Kant, not only inevitably leads to the idea of God, but which also is powerful enough to stipulate and thus circumscribe His moral nature. Moreover, while Kant undoubtedly believed that moral theism fortified the foundations of moral freedom by making us think the highest good as possibly attainable, the real foundation of morality, as he emphasized, could not depend upon the concepts of God and immortality, but on freedom.[23] The guarantee of freedom lay in the act of judgment in which people may prescribe moral laws, make unconditional demands, and by so doing set themselves ideal standards to which they may aspire.[24]

The principle used to formulate our ideals is of course, for Kant, the categorical imperative. Given this imperative is formulated in a number of different ways by Kant, and given it is one of the most discussed ideas in philosophy, it suffices to focus upon the main purpose of the imperative.

Firstly the imperative is universal and unconditional in its demands. A categorical imperative is binding on all, oneself included. In formulating a categorical imperative one asks oneself to universalize the motive of the intended action. Kant calls the following formulation of the categorical imperative the '*Fundamental Law of Pure Practical Reason*: So act that the maxim of your will could always hold at the same time as a principle establishing universal law.'[25] As a principle it is meant to supply the ideal which a person holds to be worthy as an ideal for all rational beings.

The categorical imperative supplies the form of the moral judgment. The material of the moral judgment must comply with the the form of the judgment, and not vice-versa. (This is of course the same procedure Kant employs for stipulating the form of judgments of experience). But this does not mean that the categorical imperative provides the variables of an ethical judgment.[26] Thus for example if we ask ourselves whether killing is an unconditional command, we are asking a different type of question than whether it exists. The imperative, to repeat, has the function of providing an ideal standard. Thus in the case of killing, for example, the question is not whether it is convenient to kill some people, nor whether it will make us happy if we kill some people, but whether all should agree that killing is a moral obligation.[27] In this respect the categorical imperative 'do not kill' is to be separated from any prudential commands, or what Kant calls hypothetical imperatives. These latter imperatives stipulate rules for reaching desired ends. In Kant's terminology they are technical not practical, thus they bear a closer affinity with natural science than with morality; they are governed by necessity not freedom.

Secondly, the categorical imperative requires that the person formulating the law is not only legislator and subject to the law, but that the person, as a rational being, is the end of the law. In other words, the categorical imperative requires that all people be treated as ends, never as means. The law thus requires mutual respect for persons, one's own person included.

Thus far we have emphasized the purely rational side of the moral imperative. But the rational form (i.e. the unconditionality of the imperative itself), the rational end (the person) and the rational ground (the autonomy of the will) of the judgment do relate to a material content, to things in nature. In other words while nature cannot be considered the ground of morality, in the fulfilment of duty there is an attempt to make nature conform to what is willed, and to what is rationally desirable. Thus there has to be a way of applying a pure rule of freedom to nature. And indeed this is the real centre of Kant's moral problem. The same type of problem was faced by Kant in the legitimation of the rules of theoretical reason.

In the exercise of theoretical reason the imagination held, for Kant, the key to the solution. But in a practical judgment we do not need any help from the imagination to legitimate what is moral. The morally good is merely an idea of reason, it is not an object of possible intuition. It is an archetype, not a model which the imagination constructs in order to replicate a fact. There is thus no 'schema' needed to represent the morally good in *concreto*. How then does the moral law apply to nature? Kant answers that we think the idea of the moral law in conjunction with the form of nature. The form of nature is the form of legality, and the 'legality of nature', as Kant argued in the first *Critique*, is supplied by consciousness, more specifically the understanding. In making moral judgments we make a general assumption that the ethical law *should* and *could*, if nature were to conform to reason, be a law of nature. In this respect the form of the moral law and the form of nature are conceived of as overlapping, as if intelligence could supply the ground of nature.[28] The form of nature when used in conjunction with the moral law is called by Kant the 'typic of moral law'. The 'typic' governs the formulation of the moral law in conjunction with nature.

> Ask yourself whether, if the action which you propose should take place by a law of nature of which you yourself were a part, you could regard it as possible through your will.[29]

The point of the 'typic' is that in forming a maxim one must ask oneself if this were to become a universal law of nature 'would one assent of one's own will to being a member of such an order of things?'[30]

In some cases Kant believes that even in desiring the end, it would be a contradiction to will that your maxim become a universal law of nature.[31] Kant gives the example of deception in order to gain a loan which one knows one will not repay.[32] Kant observes that in universalising the maxim it immediately becomes apparent that I do not wish all people to adopt this formula. This is not (as some critics mistakenly claim) because Kant upholds the value of property above the welfare of human beings, nor because he was uncritically bound to social practices operating in his world (though he may well have been), but because if the maxim were universally adopted it would be self-defeating. In deceiving I want the other person to believe that I am sincere in making a promise, so that I can gain by breaching the promise. I do not wish to do away with promises. On the contrary I recognize their value, I just want to gain a particular advantage by other people sticking to the rule as I deviate from it. The moral is recognized in the breach.[33] The 'typic', then, is the capacity to universalize one's intention and to ask oneself what would the world be like if it conformed to the intention. In asking this, the person making the judgment is required to think a unity of the particular (the person) and the universal (the 'realm' of rationality encompassing the idea of humanity in its rational aspect). One must consider the principle motivating action, not only from the perspective of oneself, one's ancestors or one's immediate associates, but from the position of the universal.

There is, of course, no guarantee that all members of different nations, nor members within the same nation, the same class, the same neighbourhood, even the same person in the course of his or her life etc., will agree about the content of morality. Kant's categorical imperative no more provides — nor is meant to provide — a rule guaranteeing the homogeneity of thought than do 'principles of pure reason' create laws of nature. Rather the imperative seeks to secure the space in which legitimation of one's values takes place, by specifying the rules governing that space. Within that space the person must make a decision in which humanity, as a rational, hence archetypical subject, should be considered. Other voices can quickly respond if there is disagreement. But each expression of disagreement, if it is to have any legitimacy, is an attempt to provide a sharper formulation of a universal principle. Once again the activity is analogous with Kant's understanding of the spirit of scientific inquiry as an endless progression in which one attempts to define ever more complex rules with ever greater precision. Just as scientific thought is a cooperative enterprise, moral judgments, for Kant, involve one in a cooperative enterprise, cooperation with the rational aspect of humanity. But in the moment of judgment there is also a necessary moment of particularity, a moment in which each person must answer the question 'What should I do?' The answer to that question can, for Kant, no longer simply be one of compliance with traditions, nor with one's own inclinations. The application of the categorical imperative calls these traditions and inclinations into question, not necessarily to destroy them, but to improve them. Where the inclinations or customs are grounded only in particular interests which violate human dignity and freedom then they must be transformed.

The fact that individuals or a group do not wish to enter into this dialogue or are ignorant of its possibility is no argument against Kant, just as the ignorance of geometrical formulae is no argument against them. Kant merely has attempted to establish the rules governing such a discourse. Like Rousseau he saw that the discourse of legitimation of intentions is common practice and that one need not be a scientist, an historian or a philosopher to engage in it. Indeed, again like Rousseau, he believed that philosophers, by emphasising the differences in behaviour between peoples and by seeking to define the good as something extrinsic to the will (and thus extrinsic to the judgment of legitimacy itself), often fall behind the 'common moral understanding'.

The status of the terms which govern moral discourse — the free will, the person as a rational being and end in itself, and the moral imperative as the unconditioned ideal — are for Kant not concepts that can be exhibited in space and time. They are, in Kant's terms, 'mere ideas', *noumena*, existing nowhere apart from the practical reason of the will. An empirical inquiry cannot confirm the existence of the type of concepts they are. Empiricism inevitably substitutes factual concepts (e.g. physiological/anthropological/sociological/historical) for them. If empiricism could show that the moral ideas were false (and not merely inaccessible to its procedure) then, for Kant, the discourse would be illegitimate and science would rightly require its abandonment. The Hobbesian/Spinozian world-view of struggling powers which Rousseau was countering (and which was to be reiterated by Nietzsche) would be apposite. But by demonstrating the different scope of judgments of experience and judgments of moral autonomy, Kant believes he has prevented the illegitimate encroachment of empirical science onto the terrain of morality (and, later, aesthetics). Science must remain within its own domain. In that domain judgments can be true or false, and the legitimacy of the content of a specific judgment is confirmed or refuted by experience. However, in the domain of moral ideas, the legitimacy or illegitimacy of the moral judgment is dependent upon a mere principle of reason (the categorical imperative). The principle although rational, i.e. the product of reason's own making, is not capable of being classified as true or false. Freedom, as well as the theistic concepts which it sustains, is, for Kant, a matter of a rational faith.

The new science had, for Kant, opened up one side of Enlightenment, but it could not answer the question: 'what ends should humans pursue?' By following Rousseau, by reworking the idea of the general will into the categorical imperative, and then specifying the site and status of moral ideas, Kant believed he had clearly formulated the other side of Enlightenment. A rational faith in freedom was the necessary complement of scientific activity.

In Kant, metaphysics, science and freedom harmonize in a philosophical system that has perhaps never been equalled. But there were questions this great system left unanswered, questions that were directed at his metaphysics, and which spread out into questions of science and freedom. In rapid succession Fichte, Schelling, and finally Hegel opened up a new discourse originating within the gaps (the omissions and errors) they found in Kant's critical philosophy. Kant saw the beginnings of this debate. He believed that Fichte and Schelling were merely reviving rationalism.[34] Yet there was one all important difference. If this new discourse was a revival of rationalism, it was a rationalism inevitably structured as a response to Kant's own philosophy, particularly the 'Transcendental Deduction' (in the age of Fichte) and the *Critique of Judgment* (in the case of Schelling). These new questions and the shift in the philosophical terrain are now to be examined.

Notes

1. Kant saw Rousseau as the Newton of the moral world. Just as Newton discovered the law which governed all heavenly bodies, Rousseau, for Kant, had discovered the hidden law of moral experience. *Gesammelte Schriften*, Vol. 20, p. 58.
2. Rousseau's philosophical influence in Germany was as much upon forefathers of Romanticism such as Jacobi as it was upon Kant. I am not attempting to reconcile the complex and possibly irreconcilable strands of Rousseau. Rather I am focussing upon those ideas which are fundamental for understanding Kant's dualism between necessity and freedom.
3. In *A Discourse on the Moral Effects of the Arts and Sciences* Rousseau draws a connection between the new epistemology, new metaphysics and the conception of human beings as self-interested beasts. He writes of the new philosophers: 'One of them teaches that there is no such thing as matter, but that everything exists only in representation. Another declares that there is no other substance than matter, and no other God than the world itself. A third tells you that there are no such things as virtues and vices, and that moral good and evil are chimeras; while a fourth informs you that men are only beasts of prey, and may conscientiously devour one another'. *The Social Contract and Discourses*, tr. G. D. H. Cole, (London: J. M. Dent, 1973 [1913]), p. 23. In 'The Creed of a Savoyard Priest' Rousseau defends the principles of justice and virtue as universal against the 'so-called wise men' who say that 'there is nothing in the human mind but what is gained by experience; and we judge everything by means of the ideas we have acquired. They go further; they even venture to reject the clear and universal agreement of all peoples, and to set against this striking unanimity in the judgment of mankind, they seek out some obscure

exception known to themselves alone; as if the whole trend of nature were rendered null by the depravity of a single nation, and as if the existence of monstrosities made an end to species'. *Émile*, tr. Barbara Foxley, (London: J. M. Dent and Sons 1974, [1911]), p. 252.

4. Arguing against Grotius, Rousseau says 'His usual method of reasoning is constantly to establish right by fact. It would be possible to employ a more logical method, but none could be more favourable to tyrants.' *The Social Contract*, p. 166. Elsewhere Hobbes and Spinoza are said to spread pernicious doctrines. *A Discourse on the Moral Effects on the Arts and Sciences, ibid.*, p. 24. But this does not mean that Rousseau did not have much to learn from them.

5. *A Discourse on the Origin of Inequality*, p. 54.

6. *Reveries of the Solitary Walker*, tr. Peter France (Harmondsworth: Penguin, 1979), p. 96. Also see *The Social Contract*, pp. 177-178.

7. In *Émile* Rousseau asks: 'But how is it that the good man consents to his own hurt? Does a man go to death from self-interest? No doubt each man acts for his own good, but if there is no such thing as moral good to be taken into consideration, self-interest will only enable you to account for the deeds of the wicked'. p. 252.

8. *The Social Contract*, p. 178. The concept of the general will was previously employed by Diderot in his article on natural right in the *Encyclopedia*.

9. This is what separates Rousseau's conception of the free will from Descartes's. There is no conception of unconditional duty in Descartes. Hence in this all-important respect Descartes's conception of morality is not substantially different from Spinoza's. Neither inquires into the nature of practical freedom, i.e. the capacity to make *unconditional* moral demands. Rousseau takes this step, and Kant follows him.

10. Kant's debt to Rousseau has been noted by almost all writers on Rousseau. Ernst Cassirer's 'Kant and Rousseau' provides a well-documented account of Kant's admiration for and debt to Rousseau, as well as addressing the differences in their temperaments. *Rousseau, Kant, Goethe*, (Princeton: University Press, 1945).

11. See the second part of *On the Old Saw: That may be Right in Theory but it won't Work in Practice* which is directed against Hobbes.

12. For Kant, the search for a systematic connection between all parts of experience assumes that the universe is governed by intelligence. *K.r.V.*, B 725-726. In *Critique of Judgment*, Kant also argues that in the study of organic, self-regulating systems, there is an underlying teleological assumption involved when one is searching for the function of a part of

an organism. For Kant the teleological assumption may be formulated as a rule of scientific research: each part of a self-regulating system should be studied as if it had a purpose for the organism. In so far as purposefulness is intrinsically related to the concept of will, we are working with an idea that transcends the limits of experience. For Kant the establishing of the legitimate scope of this heuristic idea required specifying its cognitive source. That source, argued Kant, was the faculty of reflecting judgment. *K.d.U.*, para. 65, 66.

13. For Kant we may never be sure that the will is ever activated solely by a moral principle. We are not conscious of all our drives. *K.r.V.*, B 579, *Foundations of the Metaphysics of Morals*, tr. Lewis White Beck, (Indianapolis: Bobbs-Merrill), pp. 407-408. Page numbers cited are those of the Akademie edition which Beck inserts in brackets in the text and as running heads.

14. *K.r.V.*, B 578.

15. Kant's attempt to eliminate any element of appetition from the law of practical reason is reminiscent of Aristotle's definition of law in *The Politics*: 'law is intelligence without appetition.' 1287a. But, for Kant, Aristotle failed to formulate the ethical law, because, like almost all other ethical theorists apart from Rousseau, he made the moral good extrinsic to the will's autonomy. *K.p.V.*, pp. 62-64.

16. See esp. *K.d.U.*, pp. 388-389, *K.p.V.*, pp. 25-26, *Foundations of the Metaphysics of Morals*, pp. 399, 418.

17. Personal sacrifice pervades Kant's conception of ethics. Near the beginning of the *Foundations of the Metaphysics of Morals*, p. 398, Kant says that even pleasure in being kind is an insufficient ground of moral duty. Kant's insistence on duty rather than the personal well-being of the actor may seem to indicate that for Kant duty must be accompanied by misery. Schiller (partly in Kant's defense) raised this issue in *On Grace and Dignity* and Kant responded in *Religion Within the Limits of Reason Alone*, tr. Theodore Greene and Hoyt Hudson, (New York: Harper and Row,1960), pp. 18-19. The point is not that one must hate performing one's moral obligations. For Kant, any ground for performance other than the obligation itself, and this includes the pleasure that stems from performance of duty, is insufficient. The pleasure one receives in performing moral obligations may soon wear off, but moral obligation, for Kant, does not stop when one tires of the task at hand. Kant's example of Thomas More perhaps best of all illustrates what is in Kant's mind when he is thinking of a moral person: someone who will endure any hardship rather than do what one believes to be wrong. *K.p.V.*, pp. 155-6.

18. *Op.cit.*, p. 107. Like Rousseau, Kant sees poverty as the product of civilization. *K.d.U.*, para. 83. Rousseau's and Kant's deontological ethic precludes (as Aristotle's *eudaimonian* ethic does not) the ethical legitimacy of slavery. It also leads directly to the Kantian formulation that human beings should be treated as ends never as means. This formulation also differs in scope from the utilitarian principle of 'the greatest happiness of the greatest number'. For the utilitarian ethic must rely upon some other principle if it is to exclude the exploitation and suppression of minorities.
19. Immanuel Kant, *The Metaphysical Elements of Justice: Part 1 of the Metaphysics of Morals*, tr. and intro. John Ladd, (Indianapolis: Bobbs-Merrill,1965), p. 84.
20. For Kant, justice is the public application of laws grounded in morality. It requires people to behave in conformity with duties (*pflichtmässig*), rather than from duty (*aus Pflicht*). For Kant, justice is a precondition for the public expression and exercise of freedom. The importance of justice for Kant is summed up when he says 'If legal justice perishes, then it is no longer worth while for men to remain alive on this earth'. *The Metaphysical Elements of Justice*, p. 100.
21. The attempt to reconcile happiness and duty is the main problem of the 'Dialectic of Practical Reason'. The pursuit of happiness is even a moral duty for Kant, but if duty requires, then we must surrender our own happiness. Happiness is a duty not because of our nature which, to repeat, instinctively wants happiness, but because if we are unhappy we stand more chance of being tempted to commit immoral or unjust actions. *Foundations of the Metaphysics of Morals*, pp. 399, 442.
22. *K.p.V.*, p. 134.
23. Hence Kant says that the concepts of God and the soul are not conditions of the moral law, 'but only conditions of the necessary object of a will which is determined by this law.' *K.p.V.*, p. 4. They are conditions for thinking the attainability of the highest good. We oversimplify Kant's position if we believe that his rational faith is governed primarily by deeply held religious beliefs that he was not prepared to subject to criticism. He may have been wrong in supposing that the 'highest good' which is desired by rational beings must be understood in theistic terms. He may have been mistaken to ally his moral teaching so closely with tenets of theism, and no doubt the pervasive speculative elements in scientific and ethical thought played a major part in how he formulated his problem and solution of practical as well as theoretical reason. Nevertheless, the religious concepts are subordinate to an ethical teaching. Religious concepts are intended as analogues of purely

rational ideas. See *Prolegomena*, para. 57-58. The main ideas underlying Kant's moral theism are: (1) there is no guarantee that morality and happiness must be combined in this world, *even though we should strive to combine them*; (2) as natural beings we are torn between our inclinations and our duties; (3) if we were pure intelligences there would be no conflict between our desires and duties; and (4) because we are so torn holiness is beyond our reach. Just as theism in scientific inquiry provides Kant with a picture to better grasp our own finite understanding, in moral thinking it also provides him with a picture of human beings as requiring constant moral vigilance. Like Rousseau, Kant was anti-clerical. He describes clericalism 'as the constitution of a church to the extent that a *fetish-worshiper* dominates it; and this condition is always found wherever, instead of principles of morality, statutory commands, rules of faith, and observances constitute the basis and essence of the church.' *Religion Within the Limits of Reason*, pp. 167-168. He supported the confiscation of church property when the population is sufficiently enlightened. *The Metaphysical Elements of Justice*, p. 136. And he described the transcendental philosophy as 'the grave of all superstition.' *Reflexionen*, Vol. 2, p. 163.

24. Kant realized the circularity involved in the concept of moral freedom, but he saw this as no reason to abandon it. Freedom provides the rational ground of the moral law, but it is, says Kant, our recognition of the moral law that enables us to recognize our freedom. *K.p.V.*, p. 4.
25. *K.p.V.*, p. 30.
26. As Lewis White Beck points out 'the function of a formula is not to supply the variables but to provide the procedure toward a solution; it is the necessary but not the sufficient condition for a given solution.' *Studies in the Philosophy of Kant*, (Indianapolis: Bobbs-Merrill, 1965), p. 25. Elsewhere Beck likens the imperative to the rules of the syllogism. Just as the rules of logic work in conjunction with material, but nevertheless are not dependent upon specific propositions, the rules of ethics involve testing maxims, which necessarily refer to empirical concepts. See Beck's 'Introduction' to his translation of *Foundations of the Metaphysics of Morals*, (Indianapolis: Bobbs-Merrill, 1959), p. XVI.
27. Kant, in fact, supported the death penalty for murder (though he made allowance for duels and infanticide). *The Metaphysical Elements of Justice*, pp. 102-107. The basis for this penalty as in all punishments, for Kant, is in the categorical imperative. (For punishment and the categorical imperative, see *K.p.V.*, pp. 37-8.) Now, of course, few murderers after they have committed their crime would

will to be executed. But, for Kant, this misses the point of the imperative underlying justice and just punishment. The imperative wills that there be just laws. The imperative is intended to secure universal protection and liberty. Criminals do not want to forego the benefits of justice, rather they have, for whatever reason, exempted themselves momentarily from the law. (Hence, the idea of the golden rule 'do unto others what you would have them do to you' misses the point of the imperative. *Foundations of the Metaphysics of Morals*, p. 430.) In this instance, for Kant, the imperative does not revolve around the particular murderer wanting himself to be killed, rather around the murderer *qua* person wanting his or her liberty secured. Kant maintains that the death penalty is the appropriate penalty, on the basis of the unclear idea that punishment must be equal to the crime.

28. This was why Kant in the first half of *Critique of Pure Reason* persistently drew attention to the limits of phenomena and the rules and principles stipulated in the 'Transcendental Analytic'. The forms and functions of experience require that we 'read' experience in a certain way. Knowing this enables us to consider what the world would be like independently of these forms and functions. If we consider that the ground of nature is intelligence, as Kant claims we do when we see nature as purposeful, then we attempt to provide a content to what falls outside the scope of theoretical science, and judgments of facts.
29. *K.p.V.*, p. 69.
30. *K.p.V.*, p. 69.
31. Duties which conform to this condition are classified by Kant as 'imprescriptible'. *Foundations of the Metaphysics of Morals*, p. 424.
32. *Foundations of the Metaphysics of Morals*, p. 422.
33. One of the most valuable ideas in Kant's ethical theory is that it encompasses a common 'fact' of moral experience, which is often overlooked in ethical theory: honouring moral principles in the breach. To observe this fact one needs to pay close attention to the modes of moral justification and legitimation. It is beyond the scope of this work to investigate these modes, but it is worth considering a couple of revealing examples. When regimes practice torture they invariably acknowledge the value of the normative principle which they are violating. They often either deny they are practising torture or they plead that in their case there are exceptional circumstances. In Nazi Germany, in spite of the anti-semitic propaganda and practices, the Nazis took care to legitimate their activity partly by dehumanising their opponents, which is an extremely common form of legitimation, and partly by

producing propaganda films of concentration camps. In these films inmates were presented as happily spending their time doing productive activities, such as work, exercising, and studying. The Nazis knew very well what type of behaviour was generally acceptable internationally as well as at home, just as do contemporary neo-Nazis who deny that the holocaust took place. Along similar lines, it is a rather common practice for nations violating human rights to claim that they are immune from censure because they have different customs. Yet invariably the country committing the violation is a member of the United Nations, and thus has pledged to adhere to the United Nations Charter as embodied in the 'United Nations Universal Declaration of Human Rights'. Amongst other things the Declaration stipulates that 'no one be subject to torture or to cruel, inhuman or degrading treatment or punishment.' (Article 5). Here too the country recognizes the rule as it enters into a moral agreement. It hopes to gain the moral sympathies of other members of the forum who subscribe to the universal values set out in the Declaration, while making an exception for itself in particular circumstances.

34. Kant had a lengthy correspondence with Fichte. And Fichte had believed himself to be a disciple of Kant. After reading what Fichte was trying to do with the critical philosophy, Kant unequivocally distanced his project from Fichte's *Science of Knowledge*. See *Gesammelte Schriften*, Vol. 12, pp. 396-397. With Schelling, Kant may have only read a review of the *System of Transcendental Idealism*, but he grasped the association between Schelling's philosophy and Spinoza. *Gesammelte Schriften*, Vol. 21, p. 87, p. 97.

PART III
ABSOLUTE IDEALISM

1 From transcendental to absolute idealism

Anti-dualism and Kant

The context of the theoretical transition from Kant's transcendental to Hegel's absolute idealism is substantially different from the transition from Descartes to Kant. Whereas Kant was born into a Newtonian world unknown to Descartes, Kant was still alive in the formative years of Hegel's thought. Moreover Kant was well aware of the reaction against dualism expressed by Herder, Hamann and others, which was to be so pervasive in shaping Hegel's intellectual environment. The anti-dualism was not like the anti-dualism of Gassendi or Hobbes, an anti-dualism that was firmly committed to the mechanistic world and inductive procedure of resolution and composition. The new anti-dualism was a reaction against the mechanistic view of the world, a reaction against the substitution of pale, lifeless abstractions for the variegations of nature and the poetic and cultural achievements of human beings. It was an affirmation of the organic unity which is expressed in language, myth, religion, art and culture generally.[1]

In keeping with this anti-dualism was Hamann's criticism of the first *Critique*. Kant, said Hamann, had attempted to separate reason from all tradition and the shared values of the community.[2] Kant's dichotomies between the *a priori* and *a posteriori*, between pure and empirical intuition, between intuition and concepts etc. were, observed Hamann, all dependent on the received philosophical language that Kant used. Kant, according to Hamann, had committed a monumental blunder. He had described these oppositions while ignoring the very language and traditions, the underlying culture in which they arose. Herder also published a *Metakritik zur Kritik der reinen Vernunft*. Like Hamann, Herder also claimed that Kant's chief error was his failure to give primacy to the language and culture which made his *Critique* possible. Again, Kant's conceptual divisions were not definitive; they were the transitory divisions made within a greater unity, the expressions of a particular culture and language which was in a state of flux.

Kant may not have read Hamann's 'Metakritik über den Purismum der Vernunft' and although Herder's *Metakritik* was not published until 1799, he was aware of the anti-dualism orientation pervading their other work. He was also aware of the wide-spread revival of interest in Spinoza.[3]

In the third *Critique* Kant himself entered the ambit of this anti-dualism by attempting to locate the capacity we have to think a purposeful unity. In recognition of the roles played by the concepts of totality and purposive unity in artistic productions and the 'life-sciences', he sought to specify the transcendental source and thus the *a priori* scope of purposive unity which evaded the methodological procedures of mechanism. Kant saw the faculty of reflecting judgment, the capacity to conceive of a totality as purposeful, as responsible for this unity. He distinguished between aesthetic and teleological judgments. The former judgments referred to subjective purposefulness. That is, the object pleases for its own sake; to appreciate its intrinsic purposefulness does not require 'theoretical' understanding, simply a free play of the imagination and understanding. The teleological judgment, on the other hand, refers to what is objectively and conceptually purposeful, and thus of intrinsic value for scientific inquiry.

In defining the site of this unity and specifying the legitimate types of judgments which must employ the concept of finality, Kant also attempted to ascribe a type of unity to the fundamental dualism of his philosophy: the dualism of necessity (nature) and freedom (moral autonomy). Kant's synthesis between freedom and necessity remained faithful to the methodological strictures of the critical philosophy. The union between the two types of judgments remained a regulative principle, a heuristic idea.

By making mechanism 'conceivable' under the principle of teleology, and by presenting a picture of nature as a vast organic system, Kant seemed to be disclosing the underlying unity, the fundamental substrate, the *Urgrund* of the world. Yet Kant's connection between the mechanical and the organic also remains allied to the idea of the scientific supremacy of mathematical and hence mechanistic science. Kant argued that the subordination of the mechanistic to the organic exists in so far as the mechanical may be conceived as governed by a will, while a *telos* is not a necessary condition of mechanism.[4] And, Kant claimed, in order to maximize our understanding of organic forms we are required to conceive of all parts as functional and reciprocally related, *as if* they were the product of design, and hence of a will. But Kant disallowed the idea that a teleological idea can be substituted for a mechanistic concept. The former has value merely as a guide for orientation, as a means for guiding experimental research where mechanism no longer has anything definite to offer. Again, teleology is based upon a 'mere idea.' By remaining closely allied to the framework of mechanistic science and the goal of validating an abstract moral philosophy, Kant's philosophy remained unequivocally dualist in an anti-dualist philosophical environment.

The popularity of Kant's third *Critique* lay largely in the fact that the great thinker was finally addressing the problems of artistic expression, culture and nature within the context of organic unity. Significantly, the third *Critique* was enthusiastically received by the very people who were also rediscovering and celebrating Spinoza, a philosopher opposed and described by Kant as 'the true key to dogmatic metaphysics'.[5] Kant had made this judgment because Spinoza conflated freedom with necessity, viewing (*Anschauung*) with conceiving, and because he had sought to define the totality in which all pieces of knowledge could be discovered (i.e. the mind of God).[6] Yet it was precisely this attempt to comprehend the unity within the world's diversity that was responsible for the rekindled interest in Spinoza.[7]

As is well documented the young Hegel entered enthusiastically into this anti-dualist milieu in which Spinoza's austere rationalism converged with elements of Romanticism.[8] We hear the echo of Herder, Hamann, Schiller, Hölderlin and others in Hegel's mature view that religion, mythology, art, language and the state embody the essential spirit of a people. Yet the philosophical reasoning behind Hegel's 'absolute idealism' is most clear when viewed as a critical synthesis of Fichte's engagement with Kant's critical philosophy, and Schelling's response to both Fichte and Kant.[9]

The new metalevel: Fichte and the conditions of the transcendental conditions

Fichte claimed that he was unable to say anything not already stated 'directly or indirectly, and with more or less clarity' by Kant.[10] Yet the question he posed to the critical philosophy and the answer he provided demonstrate that instead of rounding off the critical system, as he claimed, his philosophy played a large part in shifting the philosophical agenda in Germany away from that of the critical philosophy.

According to Fichte, Kant's great intellectual step lay in the positing of the transcendental ego as the source of the conditions of knowledge. But Kant, he observes, instead of demonstrating that the conditions of consciousness are themselves necessary and in need of theoretical justification, merely sets out these conditions as if they were proven. Kant had to have a knowledge of the transcendental *a priori* conditions (the forms and functions of experience) in order to have a *Critique*, but how can he have such a knowledge? This too must be based upon conditions. A complete system would not leave uninspected any conditions of any type of knowledge, including the knowledge required to undertake a 'transcendental critique.' Fichte argues that Kant must have relied upon conditions of consciousness in order to make a transcendental critique.

> I am aware that he by no means *proved* the categories he set up to be conditions of self-consciousness, but merely said that they were so: that still less did he derive space and time as conditions thereof, or that which is *inseparable* from them in the original consciousness and fills them both; in that he never once says of them, as he expressly does of the categories, that they are such conditions, but merely implies it by way of the argument given above. However, I think I also know with equal certainty that *Kant envisaged* such a system; that everything that he actually propounds consists of fragments and consequences of such a system, and that his claims have sense and consequence only on this assumption.[11]

Kant had not posed the question: 'how is it possible that consciousness observes the acts of consciousness?' The absolute foundation is touched, says Fichte, but it is skipped over. Brushing aside the entire methodology of the 'transcendental critique', Fichte holds that Kant should have deduced all conditions of knowledge from the one overarching principle of consciousness which is at the basis of the *Critique of Pure Reason*, the act of transcendental apperception. Fichte equates the act of apperception with the act of intellectual intuition and makes this the necessary point at which consciousness observes the acts of consciousness. Only by virtue of this faculty, claims Fichte, am I immediately conscious of the fact that I act and how I act. There is no need to derive this faculty out of any other concepts. Without it no knowledge is possible. For Fichte there is as little point in attempting to derive this original act from other concepts as there is in attempting to explain colour to one who is born blind.[12] The act is at once an act of self-postulation and the demonstration of freedom, the seat of theoretical and practical reason. The postulation and freedom are specific 'acts' (*Thathandlungen*) of consciousness. The task of philosophy is to specify the structure of these 'acts', to show their derivation from consciousness, to provide 'a genetic derivation of that which occurs in consciousness.'[13] Fichte calls the science that undertakes this task the 'Science of Knowledge' (*Wissenschaftslehre*). It is meant to supply the ground of *all the sciences*, including Kant's, and Fichte's own philosophical preconditions.

> The science of knowledge should, however, not only give the form of *itself*, but it should give the form of *all possible sciences*, and establish with certainty the validity of this form for all sciences. This can only be conceived under the condition that everything which is supposed to be a proposition in a science should already be contained in a proposition within the science of knowledge, and that the proposition is already set out in its appropriate form.[14]

Fichte will say in Kantian fashion 'Philosophy...must...furnish the ground of all experience.'[15] But, in spite of the similarity in expression, Fichte remains much closer in spirit to the anti-dualist attempt to specify the unity within diversity than to Kant. This is evident in a number of ways. Firstly, Fichte's task is to display the necessary unity between the form and the content of science. 'In the science of knowledge the form is never separate from the content, nor the content from the form; in each of its propositions both are intimately united.'[16] Secondly, consciousness is both free and governed by strict necessity. It is free in so far as it lies beyond the empirical world, the not-I. It is governed by necessity in so far as its 'acts' are necessary. The spheres of the 'I' and 'not-I', consciousness and the 'picture' of the world are never totally separated; they are two spheres with the same content. The limits of intelligence, and thus the dynamic of the sciences are not determined from outside us, but by the development of consciousness itself. The intellect, says Fichte, 'does not register some external impression, but feels in this action the limits of its own being.'[17]

Not surprisingly the basic opposition between the *a priori* and the *a posteriori* which is responsible for Kant's entire problem are merged by Fichte who declares:

> For a completed idealism the a priori and the a posteriori are by no means twofold, but perfectly unitary; they are merely two points of view, to be distinguished solely by the mode of our approach.[18]

Hand in hand with his anti-dualism and his attempt to derive all conditions of knowing back to consciousness is the thoroughly pragmatic nature of Fichte's *Science of Knowledge*. Not only does Fichte describe himself as a 'pragmatic writer of history',[19] but he concedes that if any of the results of a philosophy are clearly contradicted by experience, that philosophy must certainly be false.[20] Instead of merely demonstrating, as Kant sought to do, how the forms and functions of experience are necessary in judgments of experience, Fichte has required the derivation of the conditions from a principle. Philosophy must demonstrate the genetic relationship of the different conditions. In other words Fichte is only left with two things to verify what he says: internal consistency or logic, which is itself supposedly derived from the *Science of Knowledge*, and experience, which is also supposed to be derived from the *Science of Knowledge*.[21] Kant's attempt to provide a scientific metaphysic by circumscribing the scope of *a priori* concepts and then locating their legitimate sites is thus replaced by Fichte's open-ended *Science of Knowledge*. There are always gaps for Fichte to be filled in, further determinations to be described, even though the system of consciousness is one. Fichte's idea of a system is of an expanding circle of determinations. The content of the sciences reflects the freedom of consciousness in its systematic constructions. The open-endedness that, for Kant, had characterized empirical science is transferred by Fichte to a metaphysical plane.

In spite of the above, Fichte retains, at least partially, the distinction between practical and theoretical reason. Practical reason, and thus the power of will, is primary. This may seem to reflect Kant's dictum of the primacy of practical reason. But by discarding the specific procedures of legitimation of the categories of 'theoretical reason', particularly the schematization of categories, and by eliminating the distinction between things in themselves and things, Fichte departs substantially from Kant.

From Kant's position, Fichte had merely reverted to pre-scientific metaphysics, having no safeguard other than logic against the possibility of arbitrary *a priori* associations. As Kant wrote in 'Explanation in Relation to Fichte's Science of Knowledge':

> pure science of knowledge is nothing more or less than mere logic, which, with its principles does not search for the material of cognitions, but, as pure logic, abstracts from their content, from which a real object is vainly picked out.[22]

Fichte's dynamic deductive procedure of the *Science of Knowledge* did indeed involve the substitution of a 'logic' for a metaphysics. Yet for many of Fichte's contemporaries the critical philosophy was powerless against Fichte's meta-critique. It is this position which Hegel held. In his *Lectures on The History of Philosophy* Hegel approvingly sums up Fichte's movement thus:

> If I make a pure category an object of my consciousness, so I make my consciousness an object of consciousness and so I stand behind my usual consciousness. Thus had Fichte first brought to consciousness the knowledge of knowledge (*Wissen des Wissens*).[23]

For Hegel, Fichte's greatness lay in the recognition of the need to derive the categories from a highest principle and in the recognition that the relation between categories must be logically necessary.

> The Fichtean philosophy has the great merit and importance of having laid down that philosophy must be a science from its highest principle, and all determinations are necessarily derived from this highest principle. The greatness lies in the unity of the principle and the attempt to develop in a thoroughly scientific manner the complete content of consciousness, or as someone has called it, the attempt to construct the entire world....It is the requirement of philosophy to embody a living idea. The world is a blossom eternally produced from the one seed.[24]

And along similar lines Hegel makes the following comparison between Fichte and Kant.

> Fichte does not commence as Kant does, in an explanatory manner when he begins with the ego; that is what is great in him. From the ego everything is supposed to be derived, the explanation is supposed to be sublated. — I know what lies in me; it is pure abstract knowledge, that is the ego itself. This is where Fichte begins. Kant picks up the determinations of pure knowledge, the categories, empirically; they come straight from logic — this is a completely unphilosophical and unjustifiable procedure. Fichte has advanced further, and this is his service; he has required and sought to complete the derivation, the construction of thought determinations from the ego. The ego is thinking and active, it produces its own determinations.[25]

In requiring a 'genetic derivation' of the conditions of consciousness, and in equating the world with a construction of thought determinations, Fichte was to be one of Hegel's most important precursors. He was a major link in the theoretical transition from 'transcendental' to 'dialectical logic.' But, for Hegel, there were some fundamental problems with Fichte's philosophy.

The most important problem, for Hegel, is that Fichte is unable to make his highest principle self-forming because his philosophy remains dualist, in spite of its expressed intentions to the contrary. Fichte's dualism is most evident in the necessary distinction he makes between the picture (*Bild*), with which the 'sweeping productive imagination' works, and the thing which forces itself into consciousness.[26] The I is subjected to a bump or check (*Anstoss*).[27] Fichte believed that by emphasising how the subject always moves from picture to picture, he had eliminated any vestiges of dualism. But the problem is that even though the thing which effects and is supposedly initially conditioned by an act of consciousness, the not-I, is only ever represented to consciousness through seeing, thinking, feeling etc., the not-I has a certain autonomy. The not-I becomes what it is not supposed to be, a thing in itself, which can never be known, but which consciousness strives to know. As with Kant the striving continues indefinitely.[28] Knowledge and virtue remain in Fichte as they were in Kant, the ever elusive ends of an asymptotic quest. To be sure, Fichte stresses repeatedly: (1) that the not-I is the product of the I; (2) that the I itself is feeling and suffering (*leidend*), thus the condition of the not-I; and (3) that striving is the condition of the object which is simultaneously only ever structured through consciousness. In other words, Fichte's ego is a moment in a totalizing process and his philosophy is meant to eliminate fragmentation and residual elements which lead to philosophical dualism. But this does little to change the basic problem which is as evident in the conclusion of an I restlessly striving indefinitely to attain knowledge as it is in the need for Fichte to introduce natural determinants, drives (*Triebe*), to describe the activity of the I. For Hegel, Fichte is thus caught within the same dualism as Kant.

> However Fichte's description is imprisoned from the outset by its antithesis, as in Kant we have the I, the representation and then the thing in itself; with Fichte we have the I and not I.[29]

In other words, as Hegel was to write in his early work *The Difference Between the Fichtean and Schellingian Systems of Philosophy*, any talk on Fichte's part of the identity between form and material is only abstract. The unity between oppositions is not genuinely dynamic. The I and 'not-I' are only formally combined.

> Thus the end of the system becomes unfaithful to its beginning, the result to its principle. The principle was the

I=I, the result is I= not I. The first identity is an ideal-real one in which form and matter are one; the latter is merely an ideal identity in which form and matter are separated. It is a merely formal synthesis.[30]

Hegel draws the conclusion that nature in Fichte's system is already essentially determined and thus something dead.[31] Or as he indicated in his next major work, *Faith and Knowledge*, the principles which function within the system of knowledge are not genuinely dynamic; they leave the world just as it was before we had a philosophical understanding of it.

The immediate product of this formal idealism...has...the following shape. A realm of experience without unity, a purely contingent manifold, on one side, is confronted by an empty thought on the other.[32]

For Hegel, these important defects of Fichte's philosophy were largely rectified by the philosophy he described as the 'single meaningful surpassing of the Fichtean philosophy',[33] the philosophy of his one time friend, Schelling. Schelling attempted to stipulate an identity between nature and intelligence, and thus to find a point of unity that would be genuinely unconditioned.

Schelling and the absolute

Schelling accepted the need for and legitimacy of Fichte's re-working of Kant's system. In his early work *On the Possibility of a Form of All Philosophy*, he argues, like Fichte, that it is necessary to find a highest principle out of which the transcendental elements can be derived.[34] And like Fichte, he claims that Kant did not legitimate his own procedure, 'how he himself had come to the former synthetic concepts.'[35]

For Schelling, as for Fichte before him, the capacity to construct or to organize is the basic act of philosophy. Like Fichte, Schelling sees this act as indispensable for Kant's deduction of the categories. Organisation itself relies upon the capacity to view the universal in the particular, the unity within diversity. This capacity rests, for Schelling, upon the faculty of intellectual intuition, which is also said to be the basic condition of all science. As he writes in *Further Description from the System of Philosophy*:

Intellectual intuition, not only in passing but abiding, as the unchangeable organ, is the condition of the scientific spirit generally and informs all parts of knowledge. For it is the major capacity to see the universal in the particular, the infinite in the finite, both united as a living unity. The anatomist who dissects a plant or the body of an animal, rightly believes he directly sees the individual thing, which he labels plant or body; it is only possible to see the plant in the plant, the organ in the organ and in a word, the concept or the indifference in the difference through intellectual intuition.[36]

Schelling 'freezes' consciousness at the moment of classification, the substitution of one universal for another. In this moment of assessing the adequacy of a classifier for the particular relations, we perform, according to Schelling, an act of construction. Whether it is a movement from universal to particular or from particular to universal is irrelevant. In both cases we have a rule that is immediately applied, and it is only our intuitive understanding that assesses the adequacy of the rule.

Kant had used the concept of the schema employed by the imagination to explain this capacity, but by separating the *a priori* from the *a posteriori* he had required that the legitimacy of the concept always be related back to what is viewable, and therefore testable. The only science he had allowed in which pure constructionism takes place is mathematics. Here Kant claimed constructionism was required because the universal was dependent solely upon the mind's own activity.[37] In other words the definitions of the mathematical entities were not touched by the imperfect exhibitions, the particular representations of the rule. In other sciences 'the schema' is either derived from particular empirical data, or in the case of philosophy, the activity is reflective rather than constructive. Philosophy reflects upon the source and scope of concepts. It has no privilege whatever regarding the content of the other sciences. With Schelling philosophy is required to be far more than this. As the task of philosophy is to stipulate the conditions of knowing, it must encompass the sciences as it simultaneously *constructs* that which is the ground of all knowing, the absolute or the infinite which contains all members within itself.[38] Kant had called this idea the 'transcendental ideal' corresponding to the speculative concept of God.

What for Kant was an unattainable ideal of knowledge, a maxim for systematising our thought about the world, becomes for Schelling the condition under which we think. For Schelling, the absolute is not as it was for Kant, a problematic concept. On the contrary it alone is the source of the possibility of reason. It can be known because it is the beginning and the end of knowledge. Like Fichte, Schelling is anti-dualist in his intentions.

The idealism of theoretical philosophy is thoroughly antidualist, which declares the absolute identity of subject and object in representation; if one asks what is the object, he answers I myself in my finite producing.[39]

Concomitantly, for the young Schelling, the world is not something apart from knowledge. It is knowledge. It is not something apart from the act of intellectual intuition. It is inseparable from that act. If there is anything apart from knowledge, the totality of determinations incorporated under the 'idea of the absolute', we cannot begin to discuss it without returning to the absolute principle from which all knowledge is derived.[40] Kant's dualism had been inseparable from his critique of a dogmatic metaphysics which undertook to articulate the whole, the mind of God which created and thus knew all relations stretching forwards, backwards and coexistent in time and in space. It is this very lack of dogmatism which Schelling sees as the problem. Kant, according to Schelling, has merely adopted another type of dogmatism, the dogmatism of the preference for the finite over the infinite, for the empirically immediate over system. But, for Schelling, Fichte had demonstrated that Kant's critique also relied upon conditions of knowledge. Hence Kant's successors are driven back to search for the absolute, which Kant assumes but reduces to a mere 'postulate', a mere heuristic fiction of reason. The fundamental point of Schelling's critique of Kant is encapsulated in the claim:

> We hold this for the worst effect of philosophy which is restricted to critique, that it confirms and simultaneously sanctions the dread which the servile understanding of the spirit has before the absolute.[41]

If any type of dualism is to be rejected on the basis that it fails to take account of organisation and construction as the point from which all knowledge stems, the question arises out of what material does the I organize? But this way of putting the question already involves assuming a division between nature and consciousness. This way of asking the question, for Schelling, underlies Kant's critical philosophy and its accompanying unsatisfactory solution. It may seem common sense that this division is absolute and legitimate, but Schelling's philosophy makes this division itself the product of an underlying unity, of self intuition, or what he also calls spirit (*Geist*). 'The object of intuition' writes Schelling 'is thus nothing other than the spirit itself in its activity and its suffering.'[42] Whether spirit or the absolute is represented as nature or the ego is, for Schelling, a matter of a particular starting point. One can start with one or the other, with the subject or the object. If one starts with the object one has a philosophy of nature, and if one starts with the subject, one has a 'transcendental philosophy'. Transcendental philosophy has now nothing in common with Kant but the name. Nature and the transcendental ego are now both conditioned by the absolute and the content of nature and the content of the I are interpreted as various gradations of an underlying unity. The intellectual intuition of rational beings is the point at which nature has raised itself to its goal. It is thus seen as the end of nature, the point which nature is striving to reach.

> The dead and unconscious products of nature are merely abortive attempts that she makes to reflect herself; inanimate nature so-called is actually as such an immature intelligence, so that in her phenomena the still unwitting character of intelligence is already peeping through. — Nature's highest goal, to become wholly an object to herself, is achieved only through the last and highest order of reflection, which is none other than man; or, more generally, it is what we call reason, whereby nature first completely returns into herself, and by which it becomes apparent that nature is identical from the first with what we recognize in ourselves as the intelligent and the conscious.[43]

The world is, for Schelling, a complexity of organisations, a system of intelligence striving for self-realisation. The task remaining for philosophy, for Schelling, is to trace the course of the development of intelligence. Fichte had been content to talk about the self generation of categories, but his philosophy had been only a 'formal idealism', and hence it remained one-sided.[44] A philosophy of nature (*Naturphilosophie*) is required which constructs the necessary 'hierarchies of organisation' beginning from the most elementary constructions of matter, which Schelling, following Kant, takes as expansive and attractive forces, through the more highly developed 'potencies' of electricity, and galvanism, leading to 'irritability' and the more highly organized forms of the organic world and finally to the self recognition of the alpha and omega of the process, the world soul. The system of science must reflect the system of 'spiritualisation of nature' so that the task of a *Naturphilosophie* is to show the finest rules of intelligence, those which organize matter:

> The phenomena (the matter) must wholly disappear, and only the laws (the form) remain. Hence it is, that the more lawfulness emerges in nature itself, the more the husk disappears, the phenomena themselves become more mental, and at length vanish entirely.[45]

Like Fichte, the *a priori* construction is essentially a dynamic principle of explanation which is invalidated as soon as phenomena are found that it can no longer envelop. Hence any explanatory principle of the philosophy of nature

> must still in addition be brought to an empirical test; for if all natural phenomena are not able to be derived from this postulate — indeed if in this composite of nature a single phenomenon is not necessary according to the former principle, or if it contradicts it, the postulate is necessarily already falsified, and ceases from that moment to be valid as a principle. [46]

The Kantian distinction between *a priori* and *a posteriori* has no real meaning within this context, nor does the thought of isolating different types of conditions of different types of judgments. There is simply a dynamic field of data and principles. The system is self-generating, creating new ways of observing. Only within the context of a system do data and problems hitherto unimaginable surface. When any principle fails to account for data within its horizon a new principle is to be sought for, a new truth must be born. As Schelling writes in the opening of *System of Transcendental Idealism*

> such a system finds the surest touchstone of its truth, that it not only provides a ready solution to problems hitherto insoluble, but actually generates entirely new problems, never before considered, and by a general shattering of received opinion gives rise to a new sort of truth.[47]

For Schelling the complexity of organisation does not stop with the purely theoretical act of self-reflection. It manifests itself at the practical level. Self-consciousness must manifest and organize itself as culture. History becomes a further moment in the development of the absolute as it reveals itself to itself: 'History as a whole is a progressive, gradually self-disclosing revelation of the absolute.'[48] Again reminiscent of Fichte, freedom is the beginning and end of the process. Freedom is required in the very act of theoretical construction as well as in our social existence. The social world is the embodiment of the creative and organizational powers of human beings, a tribute to the constructive ability and emancipatory energies of the species. We are, says Schelling, 'collaborators (*Mitdichter*) of the whole and [we] have ourselves invented the particular roles we play.'[49] Hence, for Schelling, the field of the mythical and the cultural must also be a major object of philosophical inquiry.

Hegel's critical assimilation of Schelling

By dealing with the content of nature, and not just the forms of consciousness, Hegel believed that the *Naturphilosophie* had eliminated the residues of dualism in Fichte. But, for Hegel, Schelling was not merely moving from the field of speculative inquiry to physics, he had demonstrated the dependence of natural science upon speculative principles. Empiricists and most scientists, claimed Hegel, do not realize how the entire field of science, and thus the content of science, is structured according to the thought determinations operative within it. Concepts such as laws and forces are, claims Hegel, 'thought determinations' which generate new ways of classifying data, new data, and new conditions. Although Kant had already attempted to demonstrate how experience is formally conditioned, Schelling had gone far beyond this by seeking to demonstrate a fundamental identity between the dynamics of thought and the dynamics of nature, thus taking much further the item of form-dependency for the study of nature.

> Physicists don't know that they think; they are just like the Englishman who discovered that he could speak prose.— Schelling's service is not that he introduced thoughts into the interpretation of nature, but rather that he transformed the categories of thinking about nature; he introduced the importance of forms of concepts, of reason into nature, as in the case of magnetism with the form of the syllogism. He not only stipulated those forms, but he also sought to develop the principle from which nature was constructed.[50]

In summing up Schelling's contribution Hegel emphasizes that Schelling had grasped the identity of subject and object. He had recognized that the highest moment of the system of knowledge, the 'idea', generates different levels of complexity of organisation until its own truth is realized as a unity of subject and object.

> The idea (*Idee*) itself is conspicuous in Schelling, that the truth is the concrete, the unity of the objective and the subjective. Each level has in the system its own form; the last is the totality of the forms. The second great point about Schelling is that in the *Naturphilosophie* he has demonstrated the forms of spirit in nature; electricity, magnetism are only external modes of the idea, of the concept. The main point about Schelling's philosophy is that it concerns itself with the content, with the truth, and grasps it as concrete. Schelling's philosophy has a deep speculative content, which is the very content that the history of philosophy has concerned itself with. Thought is

free for itself, but not in an abstract manner; it is concrete; it comprehends itself as the world, and not just as an intellectual world, but as the intellectual-actual world. The intellectual world is the truth of nature, of nature as it really is. Schelling comprehended this concrete content.[51]

As this citation indicates, Hegel's debt to Schelling cannot be over-emphasized. But their respective answers to the question 'how does one know the absolute?' involved a necessary break between Hegel and Schelling.

For Schelling, the answer to this question was bound up with the act of intellectual intuition which lies at the basis of his system. The truth of the absolute is, for Schelling, not essentially discursive, because it must be grasped in an act of intellectual intuition. As Schelling writes in *System of Transcendental Idealism*:

> An absolutely simple and identical cannot be grasped or communicated through description, nor through concepts at all. It can only be intuited. Such an intuition is the organ of all philosophy.— But this intuition...is an intellectual rather than a sensory one, and has as its object neither the objective nor the subjective, but the absolutely identical.[52]

Following Kant and Schiller, Schelling saw the grasping of the parts within the whole as essentially an aesthetic act. Hence the privileging of the faculty of intellectual intuition has as its corollary the privileging of the aesthetic act as the highest act of self-consciousness. Art is even superior to philosophy as a vehicle of the absolute's self-disclosure:

> Philosophy attains, indeed, to the highest, but it brings man to this summit only so to say, the fraction of a man. Art brings the whole man, as he is, to that point, namely to a knowledge of the highest, and this is what underlies the eternal difference and the marvel of art.[53]

It is from poetry that philosophy has arisen, and it is to poetry that philosophy must return if it is to guide the species to its proper destiny. Schelling's philosophy, as with Nietzsche's and Heidegger's after him, is a preparatory philosophy to a new poetry, a mythology of the future, 'the creation, not of some individual author, but of a new race, personifying, as it were, one single poet.'[54]

The elevation of the faculty of intellectual intuition and the creative power coincides with a metaphysics in which will is a condition of being; it is will which generates existence. As he was to write in *On the Essence of Human Freedom*: 'In the final and highest instance there is no other Being than Will. Will is primordial Being, and all predicates apply to it alone.'[55] Even the 'divine understanding' is dependent upon the 'divine will'. 'In the divine understanding there is a system; God himself, however, is not a system but a life.'[56] The absolute is ready to be reconstructed at any moment, not because of immanent logical contradictions within the totality, not because the system is driven by a harmonizing intelligence but because of the original creative will. This will creates order, but it is a precarious order.

> Following the eternal act of self-revelation, the world as we now behold it, is all rule, order and form; but the unruly lies ever in the depths as though it might again break through, and order and form nowhere appear to have been original, but it seems as though what had initially been unruly had been brought to order. This is the incomprehensible basis of reality in things, the irreducible remainder which cannot be resolved into the understanding (*Verstand*) by the greatest exertion but always remains in the depths. Out of this which is incomprehensible (*Verstandlosen*), understanding (*Verstand*) in the true sense is born.[57]

It is this irrational and poetic metaphysic which creates a chasm between Schelling and Hegel. For Hegel, Schelling provided no safeguard against poetic fancies and fleeting inspirations; subjective feelings could be given as much weight as genuine conceptual specifications. In Schelling's system, observes Hegel, if there is dispute about any problem, 'one can only say: if you think this is false then you do not have the intellectual intuition.'[58] This irrationalism infected the entire *Naturphilosophie*. The content of the *Naturphilosophie*, said Hegel, instead of being consistently and coherently developed out of basic conceptually defensible principles fell subject to a 'game of analogy'.[59]

Hegel never departed from Schelling's idea that the absolute is the real object of philosophical inquiry, that the absolute is real and not just a regulative principle, and that all intellectual activity relates to it. However, even when Hegel in *The Difference Between the Fichtean and Schellingian Systems of Philosophy* appeared to his contemporaries to be merely a spokesman for Schelling, Hegel had indicated that philosophy was not an aesthetic act of spontaneous organisation, but the 'speculative reconciliation' of the ostensibly discontinuous judgments of the separate sciences and intellectual activities (the 'diremptions' of the absolute).[60] If the truth is the whole, then the parts themselves must be revealed to be governed by the whole. For Hegel, this requires demonstrating the logical dynamism of the parts, specifying the mediations underlying the immediate, and how each subject is negated by its own immanent contradictions. Contradictions within the parts are themselves necessarily governed by the dynamic totality of all intelligence, the absolute. Only thus for Hegel is the real identity of subject and object established. This most fundamental identity, like every identity and every difference is, for Hegel, a product of thought. The identity between thought and being is not, as Schelling held, continually broken by the aesthetic act, rather for Hegel, it is completely determinative. Only thus could the unconditioned be really unconditioned, and the real determinant of all intelligence. The embodiment of this determinate identity is the 'concept' (*der Begriff*). The description of the absolute is the description of the development of the concept.[61] (And for Hegel, the absolute is revealed only through the entire range of mediations and contradictions that are traversed in the description.)[62] To comprehend this central task of Hegel's philosophy more fully, it is necessary to dwell upon the primacy of the 'concept' as the point of identity between subject and object. It is this idea that illumines the nature of Hegel's 'absolute idealism'.

Notes

1. Charles Taylor in chapter 1 of *Hegel*, (Cambridge: University Press, 1976) and Isaiah Berlin in 'Herder and the Enlightenment' in *Vico and Herder: Two Studies in the History of Ideas*, (London: The Hogarth Press, 1976) have both focussed upon the reasons for this anti-dualism and the connection it has with Hegel. Berlin, by exploring Herder's philosophical influences, brings out the extent of this movement which was by no means confined to Germany.
2. 'Metakritik über den Purismum der Vernunft' in *Schriften zur Sprache*, ed. Josef Simon, (Frankfurt/M: Suhrkamp, 1967), p. 222.
3. Kant, of course, knew Hamann and Herder personally. Kant's 'Reviews of Herder's *Ideas for a Philosophy of the History of Mankind*' leaves no doubt that he thinks Herder's philosophy is defective. Kant had also read Jacobi's *Letters on the Teaching of Spinoza*, as well as the Mendelssohn-Jacobi dispute. In *What is Orientation in Thinking?* and *On a Newly Raised Lofty Tone in Philosophy* Kant made it perfectly clear that the new speculative attempts in philosophy were as misguided as the old; all were forms of the 'fanaticism of reason' which threatened to destroy reason. It is also interesting to note that Kant is observing the philosophy of faith (*Glaubensphilosophie*) that Hegel will continually attack. While in *Faith and Knowledge* Hegel equates Kant, Fichte and Jacobi, for reasons we shall see later, it is Hegel, and not Kant, who attempts to revive

'speculative' philosophy. From a Kantian position, the speculative articulation of the systematic and purposive unity of the world (as opposed to the regulative theoretical or necessary practical employment of the 'idea' of a purposive unity) inevitably leads to fanaticism (*Schwärmerei*). For Kant it is largely irrelevant whether the articulation is based on feeling or on reason. In both cases one steps beyond one's finitude in attempting to articulate theoretically (i.e. objectively), and not just morally or aesthetically, what no one can know — the absolute infinite.

4. This is the key to Kant's dissolution of the antinomy of 'teleological judgment'. On the one hand, all material products and their forms must be judged as possible according to mechanical laws; on the other, some products of nature cannot be judged as possible solely on the condition of mechanical laws. See esp. *K.d.U.*, para. 70, 78.

5. *Reflexionen Kants zur kritischen Philosophie*, Vol. 2. ed. B. Erdmann, (Leipzig: Fues, 1884), No. 236. Goethe and Schelling are two prime examples of this. Both were enthused by Spinoza and Kant's third *Critique*. Inevitably, any attempt at a synthesis between Kant and Spinoza meant a substantial departure from Kant. Kant himself had learnt to his amazement that even the first *Critique* was being interpreted as endorsing Spinoza. In *What is Orientation in Thinking?* he distanced himself from Spinoza, claiming that the latter inevitably led to 'fanaticism'. Immanuel Kant, *Critique of Practical Reason and Other Writings in Moral Philosophy*, tr. Lewis White Beck, (Chicago: Uni. Press, 1949), p. 302.

6. *Reflexionen*, Vol. 2, No. 238.

7. Spinoza's rise to prominence was largely due to the public controversy between Jacobi and Mendelssohn on Lessing's relation to Spinoza. At the heart of the dispute was the question whether rationality led to nihilism and whether morality and even cognition itself draw upon an irrational faith. An excellent discussion of the 'pantheism controversy' and the rise of Spinozism in Germany is provided in chapters 2 and 3 of Frederick Beiser's *The Fate of Reason*, (Cambridge, Massachusetts: Harvard Uni. Press, 1987). Beiser's discussion of Rheinhold, Schulze and Maimon makes it an indispensable work for students of German idealism.

8. There is a vast literature on Hegel's development. H. S. Harris's *Hegel's Development: Toward the Sunlight 1770-1801*, (Oxford : Clarendon Press, 1972) gives a very thorough account of what Hegel read, what lectures he attended while at the seminary in Tübingen, as well as tracing his development in Berne and Frankfurt. Less detailed, but still of seminal importance is Wilhelm Dilthey's

Die Jugendschriften Hegels, Vol. IV, *Gesammelte Schriften*, (Leipzig: Teubner, 1925). Karl Rosenkranz's *Georg Wilhelm Friedrich Hegels Leben*, (Darmstadt: Wissenschaftliche Buchgesellschaft, rep. 1963 [1844]) is an invaluable study by a student of Hegel. Also useful is Dieter Henrich's *Hegel im Kontext*, (Frankfurt/M: Suhrkamp, 1967) particularly for the relationship between Hölderlin and Hegel (see pp. 9-40) and the theological debates in the seminary at Tübingen, 'Historische Voraussetzungen von Hegels System', pp. 41-72. Finally, a very different orientation is adopted by Georg Lukács in *The Young Hegel: Studies in the Relations between Dialectics and Economics*, tr. Rodney Livingstone, (London: Merlin, 1975). As the subtitle indicates, Lukács has a very specific set of themes he wishes to highlight.

9. In my account of Hegel's 'absolute idealism' I am not so much presenting the chronology of Hegel's thought but the major ideas that hold it together. In so far as Fichte and Schelling were highly influential in Hegel's formative years, there is necessarily some overlap between the 'logic' of his ideas and their chronology. The orientation adopted here means that I move over some of the ground covered in detail by Richard Kroner and Ernst Cassirer. Kroner's study remains the standard work on German idealism. *Von Kant bis Hegel*, (Tübingen: J. C. B. Mohr, 1961 [1921/24]). Cassirer in the third volume of *Das Erkenntnisproblem* also covered the ground from Fichte to Hegel, but unlike Kroner he drew attention to the problems within absolute idealism, particularly for the natural sciences. Although the account I provide again broadly concurs with Cassirer, as well as with Kroner, it is not a reiteration of their work.

10. 'Concerning the Concept of the *Wissenschaftslehre*, or, of So-called "Philosophy"' in *Fichte: Early Philosophical Writings*, tr. Daniel Breazeale, (Ithaca: Cornell Uni. Press, 1988), p. 30. All references from Breazeale are to the marginal numbers which refer to Immanuel Fichte's edition of the first volume of *Fichte's Werke*.

11. *Second Introduction to the Science of Knowledge, for readers who already have a philosophical system* in *Science of Knowledge with the First and Second Introductions*, tr. Peter Heath and John Lachs (Cambridge: Uni. Press 1982), p. 478. Again, the references are to the marginal numbers which refer to Immanuel Fichte's edition of the first volume of *Fichte's Werke*.

12. *Ibid.*, p. 463.
13. 'Concerning the Concept', p. 32.
14. *Ibid.*, p. 51.
15. *First Introduction to the Science of Knowledge* , *op.cit*. Heath and Lachs, p. 423.
16. 'Concerning the Concept', p. 66.

17. *First Introduction*, p. 441.
18. *Ibid.*, p. 447.
19. 'Concerning the Concept', p. 77. *Foundations of the Entire Science of Knowledge*, op.cit. Heath and Lachs, p. 222.
20. *First Introduction*, p. 447.
21. On the need to derive logic from the *Science of Knowledge*, 'Concerning the Concept', pp. 68-69.
22. *Gesammelte Schriften*, Vol. 12, p. 396.
23. *Vorlesungen über die Geschichte der Philosophie*, vol. 19, *Sämtliche Werke*, ed. Hermann Glockner, (Stuttgart: Fromann, 1959), p. 619. Unfortunately, the English translation of Hegel's *Lectures on the History of Philosophy* by Haldane and Simson is too full of errors to be of any use for scholarly purposes.
24. *Ibid.*, p. 615.
25. *Ibid.*, p. 618.
26. For the role of the picture in relationship to the 'thing' see esp. 'Outline of the Distinctive Character of the *Wissenschaftslehre* with Respect to the Theoretical Faculty', in Breazeale, p. 374 ff.
27. *Ibid.*, p. 344.
28. See e.g. *Foundations of the Entire Science of Knowledge*, in Heath and Lachs, p. 261.
29. *Geschichte der Philosophie*, p. 627.
30. *The Difference Between the Fichtean and Schellingian Systems of Philosophy*, tr. Jere Surber, (Atascadero: Ridgeview, 1978), p. 55.
31. *Ibid.*, p. 76.
32. *Faith and Knowledge*, tr. Walter Cerf and H.S. Harris, (Albany: State Uni. of New York Press, 1977), p.164
33. *Geschichte der Philosophie*, p. 646.
34. F.W.J. Schelling, *The Unconditional in Human Knowledge: Four Early Essays 1794-1796*, tr. Fritz Marti, (Lewisburg: Bucknell Uni. Press, 1980), pp.104-105. References are to marginal page numbers which refer to J.G. Cotta's edition of Schelling's *Werke*, (Stuttgart, 1856), Vol. 1. And in *On the I as Principle of Philosophy or On the Unconditional in Human Knowledge* and the *System of Transcendental Idealism*, Schelling attempts such a 'deduction' on the basis of the constructive power of the I.
35. 'Über die Konstruktion in der Philosophie' in F.W.J. Schelling, G.W.F. Hegel, *Kritisches Journal der Philosophie, 1802/1803*, (Leipzig: Reclam, 1981), pp. 188-189.
36. *Werke*, Vol. IV, p. 362.
37. *K.r.V.*, B 742-743.
38. Consequently, for Schelling there is no difference in kind between the act of mathematical and philosophical construction. All knowledge involves construction. As Schelling writes, 'There is only *one* principle of

construction, whether in philosophy or mathematics, *one* principle whereby construction occurs. For the geometer the principle is the same in all constructions and the absolute unity of space, to the philosopher it is the construction of the absolute. There is, as has already been said, only one principle that is constructed, namely, ideas (*Ideen*) and everything derived is not so much derived as constructed in its idea.' 'Über die Konstruktion in der Philosophie', p. 189. Kant's example of the comparison between geometer and philosopher is likened to giving a musician paint and a brush and then asking for a musical performance. *Ibid.*, pp. 187-188.

39. *Abhandlungen zur Erläuterung des Idealismus der Philosophie* in *Werke*, Vol. I, p. 412.
40. Schelling writes in *System of Transcendental Idealism*, 'There is no question at all of an absolute principle of *being*...what we seek is an absolute principle of *knowledge*.' And 'The transcendental philosopher does not ask what ultimate ground of our knowledge may lie *outside* the same. His question is, what is the ultimate *in our knowledge itself*, beyond which we cannot go? He seeks the principle of knowledge *within knowledge*?' tr. Peter Heath, (Charlottesville: Uni. Press of Virginia), pp. 354-355. References are to marginal numbers which refer to the page numbers of Cotta's edition of Vol. 3 of *Werke*.
41. *Fernere Darstellungen aus dem System der Philosophie* in *Werke*, Vol. 4, p. 351.
42. *Abhandlungen zur Erläuterung des Idealismus der Wissenshaftslehre*, *Werke*, Vol. I, p. 369.
43. *System of Transcendental Idealism*, p. 341.
44. Like a replay of the Fichte-Kant relationship, Schelling informed Fichte 'your knowledge is not the absolute, but a conditioned knowledge.' Letter to Fichte Oct. 3, 1801, *Gesammtaufgabe*, (Stuttgart: Fromann), ed. R. Lauth and H. Gliwitzky, Vol. 111, 5, p. 83. Prior to this Schelling had explained what he found lacking in Fichte. '*Science of Knowledge*', wrote Schelling, 'proceeds entirely in a merely logical manner; it doesn't have anything to do with reality. So far as I see it, it is the formal proof of idealism...Meanwhile I want to speak about the material proof of idealism.' Letter, Nov. 19, 1800, Vol. 111, 4, p. 363. Fichte, on the other hand, believed that the fundamental identity between I and not-I had been established in the *Science of Knowledge*, and any talk of a need for a material proof of idealism was fundamentally misplaced. Letter May 31, 1801, Vol. 111, 5, pp. 45-46. Schelling's constructionism, Fichte had observed in an earlier letter, works upon a being, nature, which in turn allows itself to be constructed through a fiction. Letter to

Schelling, Nov. 15, 1800, Vol. 111, 4, pp. 360-361. In a draft of a letter, Dec. 27, 1800, Fichte wrote that the I cannot be derived from nature. Vol. 111, 4, p. 405. The I creates a periphery (*Umkreis*) around itself, a limit which can never be transgressed. The limits of feeling and conscience (the theoretical and practical limits respectively) are the limits to which the deduction of the ego can extend. There is not a higher unity between the I and nature as Schelling claims. Vol. 111, 4, p. 405. The highest principle, the I, can only ever proceed logically (i.e. in accordance with the categories the I itself generates). What Fichte thought of the content of Schelling's *Naturphilosophie* was expressed in no uncertain terms to Jacobi. 'This fellow is in no way in agreement with himself about how and to what extent he should concede existence to nature. When he falls into the absolute, he completely disappears; it is just like mushrooms growing on the dung of his own phantasy.' March 31, 1804, Vol. 111, 5, p. 237.

45. *System of Transcendental Idealism*, pp. 340-341.
46. *Erster Entwurf eines Systems der Naturphilosophie*, Werke, Vol. 3, p. 277.
47. *System of Transcendental Idealism*, p. 330.
48. *Ibid.*, p. 603.
49. *Ibid.*, p. 602.
50. *Geschichte der Philosophie*, p. 653.
51. *Geschichte der Philosophie*, pp. 682-683.
52. *Schellings Werke*, Vol. 3, p. 625.
53. *Ibid.*, p. 623.
54. *Ibid.*, p. 629.
55. *Schelling: Of Human Freedom*, tr. James Gutmann, (Chicago: Open Court, 1936), p. 350. Page numbers refer to marginal numbers which refer to Vol 7 of *Werke*.
56. *Ibid.*, p. 399.
57. *Ibid.*, pp. 359-360. Schelling's metaphysics of will, art, and genius, albeit theologically formulated, moves along the same ontological trajectory as that covered later by Schopenhauer and Nietzsche (neither of whom seemed to have realized this.)
58. *Geschichte der Philosophie*, p. 661.
59. *Ibid.*, p. 674.
60. See *The Difference between the Fichtean and Schellingian Systems of Philosophy*, pp. 15-23. The immediate impression of this work is that Hegel is defending Schelling, because he accepts Schelling's critique of Fichte. But, like Fichte, Hegel sees the problem of diremption and reconciliation as the identity of logical oppositions, and thus as a logical problem. See pp. 25-27.
61. When Hegel was finally to present the introduction to his own philosophical system, *The Phenomenology of Spirit*, he

made it abundantly clear that he intended to distinguish himself from Schelling and his disciples. In opposition to the teacher of 'intellectual intuition' and the value of genius, Hegel holds that 'Genius, we all know, was once the rage in poetry, as it now is in philosophy; but when its productions made sense at all, such genius begat only trite prose instead of poetry, or, getting beyond that, only crazy rhetoric. So, nowadays, philosophizing by the light of nature, which regards itself as too good for the concept (*Begriff*), and as being an intuitive and poetic thinking in virtue of this deficiency, brings to market the arbitrary combination of an imagination that has only been disorganized by its thoughts, an imagery that is neither fish nor flesh, neither poetry nor philosophy.' *Phenomenology of Spirit,* tr. A.V. Miller, with minor amendments, (Oxford: Oxford Uni Press, 1977), p. 42. (In all translations of Hegel I have made modifications if I am dissatisfied with the translation. Two deviations should be mentioned here. I always use use the word concept and not Notion for *Begriff.* When *Begriff* is translated as 'Notion' an important dimension of the dialogue with Kant is lost in the translation; in translations of Kant *Begriff* is always translated as concept. The word 'Notion', especially when capitalized, also has a mysterious ring to it that is not in the German. I also use lower case for the beginning of stock Hegelian terms, to avoid giving them an emphasis which is not in the German. In other places where I make amendments to the translations of Hegel, I will cite the German in brackets.) Hegel's polemics signified the end of another philosophical friendship. Once again the pupil had turned against the master in order to perfect *the system.* To Schelling, who had questioned Hegel about these comments, Hegel had claimed that the polemics were not directed at Schelling himself but at his disciples. But a remark in Hegel's Jena notebook suggests that it was indeed Schelling's philosophy that was being attacked: 'What the *Schellingian* philosophy is in its essence will be revealed shortly. The judgment about it is already in waiting, because many already understand it.' *Theorie Werkausgabe*, Vol. 2, ed. Eva Moldenhauer and Karl Markus Michel, (Suhrkamp: Frankfurt/M, 1970), my translation, p. 548.

62. Hence Hegel's disclaimers of his prefaces. They are no substitute for the philosophical work.

2 Hegel's absolute idealism and the primacy of the concept

In one sense it is obvious that a physical thing and a concept are, as Kant insisted, different. Who would disagree with Kant that to have the word or concept of 100 Thalers is not to have the money in the pocket? But for Hegel this is an obvious and trivial distinction.[1] What, for Hegel, is far more important are the mediations which are thought within each immediate.

Now Kant well knew that no intuition means anything without a concept, that our knowledge of the sensuous is a mediated knowledge. This idea, as we saw earlier, is part of Kant's definition of the *Begriff*. But by requiring that the concept conform to the object that can be exhibited in sensuous intuition, Kant, according to Hegel, failed to grasp the nature and power of the concept.

For Hegel a concept is the sum of its expanding and necessary meanings, the essential differences which are coupled with and guided by the subject. It is a unity of identity (its universality), difference (its particularity), and the singularity in which that identity and difference are developed (its ground, *Grund*).[2] This means, for Hegel, the concept is not only itself but also its necessary negations. It is through its necessary negations that a concept is differentiated and its specificity developed. Hegel repeatedly states that were a concept merely an identity having the form of A=A it would be empty. Knowledge occurs because the concept defies the law of identity, and the law of contradiction.³ To say 'a plant is a plant', or 'science is science' is to say nothing concrete. Only when the negative and different are predicated of A do we begin to have some comprehension of it. But for the predicate to relate specifically and necessarily (and not just empirically) to that subject there must be an immanent connection, a connection which is the result of the subject's unfolding within the difference. That unfolding is dependent on what Hegel calls the dialectic. This is the '*immanent* going out', the negative movement in which difference is developed within identity.⁴ The identical is coupled with its opposite and in that coupling the two are 'sublated' (*aufgehoben*) within another concept, its ground. An elementary form of that unfolding from subject to predicate is itself, for Hegel, already prevalent in the form of identity. For in the very proposition A=A, we can see that the first A is subject, the second is predicate. We are thus able to establish a difference between the two A's. Expressing that difference we now can say A=A is equivalent to A=not A, or A=B.⁵

As this example illustrates, Hegel sees that the very form of simple identity not only contains difference, but it indicates that the dynamic nature of thought is not dependent on any specific sensuous quality of the object. Rather restlessness is a necessary condition of thought, and through its restlessness thought is continually driven to specify the difference within identity.⁶ Yet the difference is not a discrete, decentred and drifting other (for Hegel, only those who are enmeshed within a dogmatic empiricism could really believe this). The difference is the product of a logical condition.

Although the differentiation is logical, it is also historical, as is evident from the fact that the concept is of necessity a process. Nevertheless, the historical mediations are shaped by the logical contours of thought, the structured movement in the expansion of the content. Concepts are forced into continual conflict with themselves. Likewise, as the *Phenomenology* attempts to demonstrate, the shapes in which consciousness is reflected are also developed dialectically. The 'concept' is said by Hegel to be 'self-developing', guided by the logical functions and forms of syllogisms required by the very collisions inherent in the concept. Human beings are not free to choose the forms and contents of their thought, rather we are extensions of a system following its own dynamic. As Hegel says in the 'Preface' to the second edition of *Science of Logic*.

> Consequently it is much more difficult to believe that the forms of thought which permeate all our ideas — whether these are purely theoretical or contain a matter belonging to feeling, impulse, will — are means for us, rather than that we serve them, that in fact they have us in their possession;

what is there more in *us* as against them, how shall *we*, how shall *I*, set myself up as *more* universal than they, which are the universal as such?[7]

It is within the above context that Hegel holds that the sensuous cannot be the adequate measure of the truth of the concept. The content of the sensuous object, is the product of the collective and logically structured experience of intelligence. As composites of relations, all objects, vis-à-vis their specified content, are completely bound up with our knowledge, the development of our consciousness:

> in the alteration of the knowledge, the object itself alters for it too, for the knowledge that was present was essentially a knowledge of the object: as the knowledge changes, so too does the object, for it essentially belonged to this knowledge.[8]

For Hegel, any science demands the concept dependency of each immediate thing. Any genuine philosophy which claims to be a science must be idealist because it realizes the concept (and theory) dependency of whatever it considers.

> The idealism of philosophy consists in nothing else than in recognizing that the finite has no veritable being. Every philosophy is essentially an idealism or at least has idealism for its principle, and the question then is how far this principle is actually carried out. This is as true of philosophy as of religion; for religion equally does not recognize finitude as a veritable being, as something ultimate and absolute or as something underived, uncreated, eternal. Consequently the opposition of idealistic and realistic philosophy has no significance. A philosophy which ascribed veritable, ultimate, absolute being to finite existence as such, would not deserve the name of philosophy; the principles of ancient or modern philosophies, water, or matter, or atoms are *thoughts*, universals, ideal entities, not things as they immediately present themselves to us, that is, in their sensuous individuality — not even the water of Thales. For although this is also empirical water, it is at the same time also the *in itself* or *essence* of all other things, too, and these other things are not self-subsistent or grounded in themselves, but are *posited* by, are *derived* from, an *other*, from water, that is they are ideal entities.[9]

In so far as philosophical knowledge is, for Hegel, knowledge of the absolutely real, the idealism is an absolute idealism.

From Hegel's perspective, Kant's transcendental idealism, with its division between concept and object of intuition is no more capable of overleaping the idealist unity between what is for us, i.e. our knowledge, and what is in itself, i.e. the truth of what the object is, than any other philosophy.[10] All this division does, for Hegel, is omit major steps in the process of knowing; it isolates the result of a complex series of mediations from the very process that made it possible. For Hegel, this omission is at the very heart of Kant's philosophy.[11] It is based on the primacy of reflection and isolation over the higher unity of reason. For the very faculties which Kant takes as the source of the *a priori* elements are themselves dependent upon the state of our knowledge of cognition (psychology) and logic.[12] Thus, for Hegel, the transcendental logic is not determinant but determined by the 'knowledges' it is supposed to underpin. Hegel's criticism of Kant, his critical assimilation of Fichte and Schelling and the resultant 'absolute idealism' are, however, not solely motivated by the disinterested probings or problem solving of the ontologist and epistemologist. Rather they are linked with a far broader critical project undertaken by Hegel which we must now consider.

Notes

1. *Logic*, tr. William Wallace, (Oxford: Clarendon Press, 1975), para. 51, where Hegel refers to 'the trivial observation (*die triviale Bemerkung*) of the *Critique*' that 'thought and being are different.' And he says the definition of the finite consists only in its existential presence (*Dasein*) being separated from its concept. But such a separation is one of reflection, one that does not take into account the process of mediation. For Hegel, Kant's claim that the concept of 100 Thalers has not one Thaler more nor less than a 100 is 'barbaric', because the content of a thing is the totality of its relations. And one of the fundamental requirements for a Thaler to be a Thaler and not just an imagined entity is that it can be exchanged, i.e. that it is not just something in one's head. See also *Science of Logic Theorie*, pp. 87-88.
2. 'Universality, particularity, and individuality are, taken in the abstract, the same as identity, difference, and ground. But the universal is the self-identical, with the express qualification, that it simultaneously contains the particular and the individual. Again, the particular is the different or the specific character, but with the qualification that it is in itself universal and is as an individual. Similarly the individual must be understood to be a subject or substratum, which involves the genus and species in itself and possesses a substantial existence.' *Logic*, para. 164.

3. This too is a persistent theme of Hegel. But see especially *Science of Logic*, pp. 413-416.
4. *Logic*, para. 81.
5. This point is clearly put in *The Difference Between the Fichtean and Schellingian Systems of Philosophy*, pp. 24-27.
6. Hence Hegel sees the law of excluded middle as needing to be substituted by the affirmation of contradiction as expressed in the sentence: 'everything is inherently contradictory.' *Science of Logic*, p. 439.
7. *Science of Logic*, p. 35.
8. *The Phenomenology of Spirit*, p. 54.
9. *Science of Logic*, pp. 154-155.
10. For the conceptual relationship between knowledge as for us, and truth as in itself see *Phenomenology*, pp. 14-16.
11. Hegel recognizes that Kant had provided a point of union between thought and intuition in the faculty of the 'transcendental imagination', and he made the 'transcendental act of apperception' the condition of all consciousness. But Hegel objects to Kant's stipulation that reason is objectively content-forming only when it creates practical/moral principles. Otherwise for Kant, the content of reason's own making is only regulative, and thus not objectively real. For Hegel this means that Kant's philosophy remains unable to account for its knowledge of the very unity which it exercises in its initial separation of the faculties, the elements of knowledge, and the types of judgment. Thus, as we saw earlier, Hegel accepts Fichte's critique of Kant. Hegel believed that only in the third *Critique* does Kant attempt a resolution. But here also the point of resolution remains either aesthetic (i.e. we have a feeling of sublimity), or if logical, as a heuristic maxim. In either case its reality cannot be confirmed. The highest point of union remains ever beyond (*jenseits*), an ostensibly unknowable thing in itself. Yet, observes Hegel, there is nothing easier to know than the 'thing in itself'. (*Logic*, para. 44). Obviously if it is not a phenomenon, if it does not conform to the principles of the understanding, if it is a border concept, etc., we *know* quite a bit about it. For Hegel it is an abstract product of reason. Kant's philosophy, for Hegel, is thus caught within the contradiction of employing a point of unity, as it then declares not only the separation between *Erscheinungen* and the thing in itself, but that one can only have knowledge of *Erscheinungen*. This latest declaration assumes a type of knowledge (i.e. empirical knowledge) as the only legitimate type of objective knowledge. Moreover, the theory contradicts the very restrictions it imposes. Or, to use another recurrent formulation of Hegel, it fails to account for the fact that in postulating the barrier to reason, one has already moved

beyond it. (For a succinct and clear formulation of this often repeated point, see *Fragmente zur Philosophie des Geistes (1822)*, 'Zweites Brüchstück', *Theorie Werkausgabe*, ed. Eva Moldenhauer and Karl Michel, (Suhrkamp: Frankfurt/M, 1970), Vol. 11, p. 529-530). The way around the problem (the point of union is real) had been seen, for Hegel, by Schelling. (This critique of Kant is repeated in numerous places, but see esp. the section on Kant in *Faith Knowledge*, pp. 67-96.)

12. Kant's system is only based on 'psychological and historical grounds'. *Logic*, para. 41.

3 Hegel's project: reconciling the finite with the infinite

Kant's critical philosophy is, for Hegel, symptomatic of a dualism involving an absolute separation between the infinite and the finite. This dualism, for Hegel, involved the philosophical dogmas of the primacy of the finite over the infinite, of the isolated particular over the totality, of the reflective understanding over speculative reason, of the 'dead positive' or the immediate sensuous object over the process of the concept.[1] Such thinking is seen by Hegel as eliminating a multiplicity of necessary determinants that are contained in what is supposed to be known. By so doing it devalues the power and breadth of reason or spirit (*Geist*), as it equates the contingent and the rational. What is real and valuable become increasingly dislocated from the very ground which produced their being.

One form of this separation between the finite and the infinite is, for Hegel, evident in empirical natural philosophy. Concepts such as 'matter, force, one , many, universality' as well as the syllogistic form appropriate to each observation are used in a 'completely uncritical and unconscious' way. How the structuring of science contributes to its content is left untouched.[2]

Another more dangerous exemplification of this separation between the finite and the infinite is, for Hegel, to be found in Enlightenment religious criticism. Instead of seeing religion as the concrete expression of the spirit of a people, as the logically structured historical labour of mind, and an indispensable component within the state for the education of citizens, religion is seen as mere superstition.[3] The church is but stones and wood. Such thinking fails to grasp the positive contribution that Christianity has made to the role of freedom within modern consciousness. The ideas of the infinite worth of each person, the creation of a bond of universality between all people, the abolition of slavery and the idea of the freedom of conscience have Christian origins.[4]

In sum the rigid separation between the finite and the infinite eliminates those regions of the spirit which religion embodies and passes on through education, while offering nothing concrete in their place.[5] The infinite is removed from the institutions which express its presence, and resited within the lonely confines of the heart or the conscience.[6]

Hegel sees the moral and political theory arising simultaneously with Enlightenment as illustrating this resiting of the infinite. The abstract rights of the person and the primacy of the individual moral conscience become the ultimate determinants in evaluating right. The absolute separation between finite and infinite is retained. Now the finite, the mechanical drives of our sensuous nature, is the absolute other which must be conquered by the infinite abstract will.[7] The social shapes which these drives assume and which contribute to ethical life are ignored as pure practical reason strives to shape the world in accordance with an unattainable idea of perfection. In either case, for Hegel, within the culture of Enlightenment the finite (the empirical data) and the infinite (the 'idea') appear as unconnected oppositions.

For Hegel there is no denying that the movement of Enlightenment and the growth of empiricism were important and valuable steps of intellectual development. However, there is an urgent need to comprehend the totality in which this consciousness operates. Enlightenment must become enlightened about itself. The age must be made aware of the uncritical assumptions embedded in its elimination of the vast regions of the spirit that are neither immediate nor sensuous. On the one hand it must be made aware of the dogma of the 'faith' in the primacy of the sensuous and the finite. On the other, it must be made aware of its separation of the abstract moral and infinite will from its social context and its being placed beyond the grasp of empiricism. According to Hegel, the effect of this thinking extends to every sphere of theoretical and practical activity. Its social expression is to be found in the event which pervades Hegel's entire philosophy — the French revolution, and its accompanying reign of Terror.[8]

In so far as Hegel's task involves a refocussing upon the unity which actively determines the oppositions, the actual infinite and the ground of all being, this required an acknowledgment of the value of the intellectual orientation of religion. As Hegel states in his *Lectures on the Philosophy of Religion*:

> The religious consciousness...is in itself the *departure* from and *leaving behind* of the immediate, the *finite* and the transition to the intellectual or, objectively determined, the gathering of the past in its absolute, substantial essence. Religion is the consciousness of the true in and for itself in opposition to the sensuous, finite truth and perception. It is consequently elevation, reflection, the transition from the immediate, sensuous, particular.[9]

As important as religion is for Hegel, religion, like art, is seen as expressing the truth in a less scientific form than philosophy.[10] The medium of religion is not the concept, but the feelings, the sensory image and the unanalyzed universal, i.e. the ineffable, the one, etc. This same truth is also, for Hegel, expressed in the great speculative systems of philosophy and the arguments for God's existence, which Hegel revives and defends against Kant's criticisms.[11] In the 'Preface' to *The Phenomenology of Spirit* Hegel states his orientation thus:

> The eye of the spirit had to be forcibly turned and held fast to the things of this world; and it has taken a long time before the lucidity which only heavenly things used to have could penetrate the dullness and confusion in which the sense of worldly things was enveloped, and so make attention to the here and now as such, attention to what has been called 'experience', an interesting and valid enterprise. Now we seem to need just the opposite: sense is so fast rooted in earthly things that it requires just as much force to raise it. The spirit shows itself as so impoverished that, like a wanderer in the desert craving for a mere mouthful of water, it seems to crave for its refreshment only the bare feeling of the divine in general. By the little which now satisfies spirit, we can now measure its loss.[12]

The final two sentences of this passage refer to the critique which runs parallel with the critique of Enlightenment: the critique of Romanticism. The age of Enlightenment brings its other, its anti-Enlightenment philosophy with it. This latter philosophy in aspiring after the infinite is unable to reach it with the methodological strictures required by empiricism. It circumvents the restrictions of empirical science by grasping the infinite in the irrational moment of 'rational faith' or 'feeling'.

All of Hegel's major works are pervaded by his constant polemic with this irrationalism. In *Faith and Knowledge* Kant,

Jacobi and Fichte are placed within this camp. Kant is placed here because he retains the absolute as unknowable yet necessary, and thus as a postulate of reason. To this extent Kant embodies, for Hegel, the contradictory sides of Enlightenment: its simultaneous abandonment of the absolute as real and its retention as an idea or moral ideal. Jacobi is placed here because he argues that if our perceptions are not only conditioned but dependent upon immediate knowledge, so too must our knowledge of the unconditioned. Hence for Jacobi the unconditioned, God, is not known mediately, through reasons, but immediately through faith. Fichte too, in *The Vocation of Man* rests not only knowledge but also the absolute on faith. Hegel also sees Schelling as caught up within this constellation. There are many others, including Schleiermacher, Fries and Baader who belong to the same *Gestalt*. For Hegel they all adopt a one-sided response to the *Zeitgeist*. They all embody the contradiction of the age: the contradiction between the finite and the infinite. All are united in their denial that the unconditioned is a rational object of knowledge whose presence is rationally knowable. Yet each has expressed something about it, either through an intellectual faith or genius or a religious faith of the heart.[13] They fail to see that a philosophy which is to be complete must show the process connecting these oppositions. Philosophy must be able to define the mediations which are subsumed under the ostensibly unknowable but much sought after absolute. It must attempt to *know* what these people 'feel' or 'believe' and talk about, but which is apparently *unknown* and *unknowable*.

This requires something more than art, religion or pre-Kantian speculative metaphysics could ever deliver. It is not a question of appealing to a myth, or to an ineffable unity in which all beings are said to originate. Rather the dynamic of the unity and its differentiation, the development of the absolute, which is not a only substance but also a subject,[14] has to be conceptually specified. The description of this development is not merely an historical account of contingent events but the specification of what is true and eternal, what is rational and necessary within them.[15]

Given that philosophical science must specify the dynamic of the 'spheres' of spirit or mind if it is to describe the absolute, Hegel takes as his guide the 'concretions' of mind — the concept. Hegel's philosophy is thus essentially a description of what he calls the self-development of the concept.

Within this project the privileged theoretical position Kant had made for mechanistic science is cast aside. Mathematical physics is just one more manifestation, one more *appearance* of *Geist*. This does not mean that the foundational principles of mathematics and mechanics are not important for Hegel. On the contrary, they too are the expression of the infinite within the finite and on this basis Hegel takes up the Kantian problem of demonstrating the conditions of their possibility. (He attempts this in the 'Objective Logic'). But whereas Kant modifies logic to comply with the cognitive conditions of mathematics and mechanics in order to provide a transcendental logic, Hegel's dialectical logic starts from thought itself. For it is in thought itself that experience becomes a philosophical problem and something demarcated from the *a priori*. It is thought which divides between subject and object and which is able to identify their unity.

Mathematics, for Hegel, is essentially a rule based thinking, a mechanical application of definitions. Hence, for Hegel, Kant's claim that geometrical and arithmetical truths are synthetic is wrong. The sum 7+5=12 is an analytic judgment requiring no synthetic act by the mind.[16] It is because numerical and geometrical operations are analytic that a machine is capable of carrying them out.[17] It must be said, however, that Kant also sees mathematical concepts as governed by their definitions. But, as I clarified earlier, he seeks to demonstrate the ampliative nature of mathematics. And this means that there must be more than a logical analysis of definitions involved when they are combined. Hegel's failure to acknowledge this when combined with the obscurity of many of the passages of the *Science of Logic* undermines his critique of Kant.[18]

Secondly, and more importantly for the general direction of Hegel's thinking, although Hegel sees mechanics and the science of number and figure as closely connected, he sees the explication of quality as quantity as a dangerous reductionism when it is stretched too far. For what happens, says Hegel, is that those concepts that cannot be so explicated, 'freedom, law, morality, and God' are left to the individuals to give them whatever content they like.[19] This, for Hegel, is the problem with 'materialism' which wants to take one important 'stage' of the 'idea' as the absolute. In other words, to take the sphere relevant to the mathematical sciences and mechanics as the standard of knowledge is one more manifestation of the dogmatism of the age: the partial elimination of what falls beyond its purview, what it cannot grasp as sensuous, and its simultaneous restoration of the infinite as a thing of faith.

Instead of making the conditions of mechanics and mathematics constitutive of experience, and then making teleology a heuristic maxim for the observation of life forms, as Kant did, Hegel defines knowledge in a broader sense. 'Knowledge (*Erkenntnis*)' for Hegel is simply 'determining and determinate thinking.'[20] Such a definition, as well as making mathematics and mechanics merely one type of knowledge with a limited application avoids the problem faced by Kant of failing to explain the conditions of its own possibility. Note also that the above definition is empty until the precise determinations have been specified. There can be no prefabricated form which can be 'glued' onto a content. The content is revealed only in the process of knowing, and the form within science is not restricted to mechanistic experience, nor is it merely regulative in the life sciences. Form is intrinsic to each object. To repeat, all form and objects are determinations and expressions of the absolute. Concomitantly, for Hegel, all science is dependent on cultural as well as logical preconditions. Knowledge is not only theoretical, it is also social and practical. But the social and theoretical dimensions are, for Hegel, themselves mediated through mind and thus structured by necessary thought determinants.

The articulation of the genesis of that structure in its pure thought form is the task of the *Science of Logic*. These thought forms themselves emerge under particular historical conditions and within particular states. Neither the pure thought determinants, nor the societies, nor the sciences within the societies have any meaning in isolation. They are all moments of the absolute at any given moment, all members of a developing sphere of rational necessity, spirit (*Geist*). No matter which field of knowledge or practice one enters into, for Hegel, one enters into the expanding sphere of the absolute. But one enters as a vehicle of the history of the science and the world, a bearer of mediations and determinations of the concept. In this respect the real subject and object, for Hegel, as indicated earlier, is spirit. In the other sciences this, for Hegel, is not realized because the object of investigation is the result, not the process. But philosophy is not like other sciences. Its truth is the whole, the absolute as process, and thus it exemplifies absolute spirit knowing itself. Philosophy must account for the complete estrangement (*Entaüsserung*) and self return of spirit. The necessity of this estrangement must be described from the most elementary, abstract and emptiest determinant, being (*Sein*), to the most concrete determinant, 'absolute spirit'.

The description of estrangement follows a dialectical course. And from the opening move of the system Hegel demonstrates how dialectic develops within and pertains to the most elementary and abstract concept, being. When taken without any further determination, being tells us nothing. Likewise the concept nothing, taken in isolation, nevertheless 'is.' What it is is not yet specifiable, but that it is, for Hegel, is part and parcel of its *being* articulated. (Hegel is not specifying what type of beings 'being' or 'nothing' are, for this requires the employment of further determinants. The task of logic for Hegel is to demonstrate the intrinsic connections, the inherent rationality of the structures of thinking. His whole point is misconstrued if, as Feuerbach thought, we assume that he is talking about a sensuous being.) The movement between these opposites is indicative of another determinant, becoming, which is the underlying unity, and the 'speculative' resolution that the dialectic has required. 'Becoming' is itself a 'doubled' determination, depending on whether one moves from being to nothing or nothing to being, whether 'it arises or passes away.' This in turn leads to a further movement as the existent (*Dasein*) within the passing away or arising is now the object of thought.

This dialectic movement from being to nothing to becoming to *Dasein* are the first stirrings in a huge dialectically developed 'concretion' of elements and principles which constitutes Hegel's system. It advances hierarchically moving always from the least complex element, principle, level, or sphere to the more complex. Each leads to its own dissolution, as it generates a new, more complicated, constellation of explanatory concepts. Such a dissolution does not mean that the more elementary level is worthless. But its incompleteness must be acknowledged as the subject which seeks the absolute moves to a more comprehensive, yet asymmetrical, level driven on by the restlessness of the determinations of thought.

In its entirety the system is conceived as the activity of reason, i.e. the idea, in and for itself (i.e. as a system of logic), in its other (i.e. in nature), and in the union of intelligence and nature, or what Hegel calls, the returning to itself from its other.[21] This last sphere is, for Hegel, the most complex of all, the one in which intelligence recognizes itself as active and determinate, as free in the world. Each concept should be treated in the appropriate sphere, and at the appropriate level. In this respect the Hegelian system is the antithesis of a reductionist system.

The movement within each sphere is not intended to follow a temporal sequence. It is logical and dialectical, a category development, which moves toward an end, philosophy's comprehension of its own age. The purpose of philosophy is not prediction but explication of the dynamic inner unity that permeates and drives the consciousness of the age, its science and practices.

With the possible exception of Schelling's system, the nature of Hegel's system is more dynamic and its scope more extensive that anything proposed before him. There is not a major field of science whose foundations Hegel does not set forth within the *Encyclopaedia*. But the very dynamism and extension created insurmountable problems for Hegel's philosophy which are nowhere more evident than in the very area where Hegel had seen the weakness of Schelling's thought, in the *Naturphilosophie*. Now although there is much division concerning Hegel's judgments in the field of natural science,[22] there is, nevertheless, an obvious problem for Hegel's metaphysics which surfaces in attempting to describe the genesis of rational concepts and principles. I shall pause upon this as it is indicative of a deeply rooted problem pervading

Hegel's project of the reconciliation of the finite with the infinite.

The problem with the Naturphilosophie

Hegel openly states that philosophy must not only be in agreement with the empirical sciences, 'in its formation and in its development, philosophic science presupposes and is conditioned by empirical physics.'[23]   On the other hand, he is committed to a description of what is rational and eternal. The criteria here are partly supplied by the *Logic* and partly from the analysis of the metamorphosis of levels (*Stufe*) of nature. This very project not only relies upon the traditional metaphysical 'law of continuity' serving as the condition of the analysis (a spurious move which involves Hegel in rejecting the idea of evolution of species) but it was also reminiscent of Aristotle's theory of substantive forms.[24] Moreover, in so far as Hegel was committed to a metaphysics which would demonstrate the development of the infinite within the finite, he *had* to move from the 'lower', less complex level to the 'higher', more complex level by demonstrating its necessary internal contradictions. In so far as he already has empirical data, he knows where he wants to move. The only problem is how to get there, and the question then concerns the reliability of the empirical data he uses.

Hegel's philosophy is meant to demonstrate the presence of the absolute within the parts. It is not an object of faith, but a living reality divested in nature, or rather, as Hegel reminds his readers who may have forgotten the centrality of the identity of subject and object, its presence is in the 'Ways of Regarding (*Betrachtungsweisen*) Nature.'[25] Yet if the metaphysical beginning is to be the basis for the structuring of the facts, how can he have demonstrated the actuality of the presence of the absolute in nature? The identity of the subject-object breaks down as a correspondence not only must take place between empirical theory and verifiable 'facts', but also between Hegel's *Enzyclopaedia* and the empirical theory.[26] The omnipresent 'absolute' is in danger of being dispensed with as redundant if and when Hegel makes a scientific error. Or it becomes, as it was for Kant, 'a mere idea', a regulative principle which, because of our finitude, cannot be defined with certainty.   In the latter case Hegel has no more escaped the bad infinity than the thinkers he criticised.   For those working within natural science, the idea of an absolute presiding over their activity made little sense. The reasoning was clearly expressed by Croce in *What is Living and What is Dead in the Philosophy of Hegel*:

> either we think that the empirical method is capable of positing some laws, some genera, some concepts, in a word, some truths; and in that case we cannot understand why the other laws, genera, truths and concepts, the whole system of them, should not be attainable with the same method. For the activity, which posits the first naturalistic concept, reveals in that act its capacity for positing the others and the whole; just as in poetry, it is the same activity and no other which forms the first verse, and which completes the whole poem. Or else we think that the empirical method is not capable of any truth, however small; and in that case the speculative method not only has no need of the other, but can draw from it no assistance. To make verbal concessions to physics and to the empirical method is mere trifling, and satisfies nobody.[27]

Hegel did intend to use his speculative philosophy to adjudicate on scientific disputes, as in his infamous defense of Goethe's theory of colour against Newton. However, Hegel was not only defending a lost cause.[28] The way the 'scientific' judgment was advanced — through a combination of polemic, poetry and a dialectical chain of obscure classifiers developed from his *Logic* — only served to confirm the view of his contemporaries, that Hegel's *Naturphilosophie* was indistinguishable from Schelling's; although, unlike Schelling, who had considerable influence in some scientific circles, Hegel's influence never extended into this sphere.[29] Indeed, to outsiders Hegel was simply a disciple of Schelling. Those who were critical of the project were not to deem it a matter of importance that Hegel's movement between levels was based upon the 'concept' rather than 'intellectual intuition'. A change in the rule of transformation did not necessarily mean that the whole affair had ceased to be a 'game of analogies.' Even Hegel's disciples found the *Naturphilosophie* a rather barren business. As Wilhelm Dilthey wrote of Hegel's stay in Berlin between 1818 and 1831.

> He [Hegel] only held classes *Naturphilosophie* of nature, which he also described as rational physics, six times. Only in the fifth semester did he finally return to it. It was not complemented by any other lectures in the area of natural science. And none of his students followed him by doing fruitful work in this area. Where Alexander von Humboldt and then Johannes Müller ruled, there was no place for this regressive treatment of nature.[30]

* * *

Nature, for Hegel, was important because it illustrated the presence of reason in the world. Only through the presence of reason in the world was nature knowable. This same idea had been intrinsic to Descartes's metaphysics. Yet Hegel's 'absolute idealism' was fundamentally different from Cartesian dualism in its problematization of natural science. Hegel's interest in nature was, in one important respect, the exact opposite of Descartes's: it was not primarily instrumental but speculative. Descartes and Hegel had both emphasized the absolute infinite, but, for Hegel, it was the totality of reason that had to be known not the mechanical and anatomical details.

Like Kant, Hegel's interest in nature was directly related to more metaphysical problems, particularly the problem of freedom. But whereas Kant had outlined a dualist philosophy in order to preserve the idea of freedom from nature, Hegel had sought to demonstrate the fundamental (dynamic) unity of the infinite and finite. He had attempted to illustrate that reason was not something separate or beyond (*jenseits*) the world or the age, but active within it. As freedom was an idea of reason, it was also necessary to define freedom, not, as Kant and Rousseau had done, as a standard *beyond* the world, but as present and determinant within the world. Enlightenment and the romantic response to Enlightenment had, for Hegel, both contributed to the obfuscation of freedom in an age in which freedom had become the central philosophical problem. Hegel's task of reconciling the infinite and the finite and describing the presence of *Geist* within his age necessarily involved articulating the rational shape of 'real freedom'.

Notes

1. This central theme of Hegel's has its first clearly developed philosophically in *Faith and Knowledge*. The importance of this work for an understanding of Hegel's thought cannot be overestimated.
2. *Logic*, para. 38.
3. For Hegel 'religion is the sphere in which a nation gives itself the definition of what it regards as the true. A definition contains everything that belongs to the essence of an object; reducing its nature to its simple characteristic predicate, as a mirror for every predicate — the generic soul pervading all its details. The conception of God, therefore, constitutes the basis of a people's character.' *The Philosophy of History*, tr. J. Sibree, (New York: Dover, 1956), p. 50.
4. This idea of Christianity is recurrent in Hegel. It is succinctly put in para. 163 of *Logic*.

5. In *The Phenomenology of Spirit* Hegel writes: 'If all prejudice and superstition have been banished, the question arises, *What next? What is the truth Enlightenment has propogated in their stead?* ...In its approach to what, for faith, is absolute spirit, it interprets any *determinateness* it discovers there as wood, stone, etc., as particular, real things. Since in this way it grasps in general *every determinateness*, i.e. all content, as something *finite*, as a *human entity* and [*mere*] *idea*, absolute being becomes for it a *vacuum* to which no determinations, no predicates can be attributed.' p. 340.
6. This may seem to indicate that Hegel's religious sentiments were Catholic, like so many romantics who converted (Schlegel, Görres, Müller). But Hegel insisted on remaining a Lutheran. In a letter to Niethammer (10 Oct. 1816) Hegel succinctly expressed the rationale for his choice of denomination: 'The Catholic community has in its hierarchy a fixed centre which the Protestant lacks. Moreover, in the former everything depends on how the clergy is instructed, whereas in the latter the instruction of the laity has equal importance. For we really do not have a laity, since *all* members of the community have the same right and role in the determination and preservation of church affairs in doctrine and discipline. Our safeguard is thus not the aggregate of council pronouncements, nor a clergy empowered to preserve such pronouncements, but is rather only the collective culture of the community. Our more immediate safeguard is thus the universities and institutions for general culture.' *Hegel: The Letters,* tr. Clark Butler and Christiane Seiler, (Bloomington: Indiana Uni. Press, 1984), p. 328.
7. This critique is fundamental to Hegel's *Philosophy of Right* and his critique of Rousseau, Kant and Fichte. It is to be dealt with more fully below.
8. The dynamic from Enlightenment to Terror is, of course, a major section in the *Phenomenology*.
9. *Theorie Werkausgabe*, Vol. 16, my translation, p. 106
10. Hence in the sphere of 'absolute spirit' art and revealed religion are subordinate to philosophy.
11. This is a recurrent theme in Hegel's engagement with Kant, but it receives its most prolonged attention in 'Lectures on the Proof of the Existence of God'. Parmenides, in the ancient world, and Spinoza, in the modern, are, for Hegel, the most important philosophical monists. Both, for Hegel, suffer from the problem that the determinant principles of the 'one' are not adequately defined. Hence the 'one' appears to be an immobile and abstract one, an 'eternal night'. See e.g. *Theorie Werkausgabe*, Vol. 17, p. 494. For Hegel, Heraclitus provides the major necessary

complementation to Parmenides, and Fichte and Schelling provide it for Spinoza. In spite of Hegel's criticism of Spinoza's system as rigid and formalistic, he holds Spinoza in particularly high esteem for his recognition that the actual infinite is not equivalent to a summation of the parts. This idea is of fundamental importance to Hegel's critique of reductive empiricism. Spinoza uses the example of two non-concentric circles in which one encloses the other. He points out that 'no number can express the inequalities of the distance which exists between the two circles, nor all the variations which matter in motion in the intervening space may undergo.' Letter to Lewis Meyer [XXIX] in *Works of Spinoza*, Vol. 2, p. 321. The same idea is exemplified by the fraction 2/7. Neither part has a meaning in itself. The same relation can be expressed as 4/14 etc. Nor can the relations be fully expressed as a decimal; the decimal is always striving yet failing to realise the totality. In *Science of Logic* Hegel contrasts the idea of Spinoza that our knowledge consists in the recognition of the underlying unity within the difference, with Kant's conception of infinity. For Hegel, Kant's entire philosophy is premised upon a 'bad infinite', an attempt to define determinants by virtue of the coalition of the parts, which themselves are capable of being split indefinitely. For Hegel's understanding of the infinite and his assessment on Kant and Spinoza on this see *Science of Logic*, pp. 240 ff. Hegel knows this is how much empirical science must progress (cf. *The Phenomenology of Spirit*, 'Perception: or the Thing and Deception'), but each progression takes place within a specific science, or intellectual sphere which provides the underlying unity.

12. *The Phenomenology of Spirit*, p. 5.
13. For Hegel, Jacobi is the key figure within this movement. In *The History of Philosophy* Jacobi's presence is said to be everywhere. 'Since Jacobi's time everything written by philosophers like Fries and theologians about God depends on this idea (*Vorstellung*) of immediate knowledge, intellectual knowledge...Everywhere one finds nothing other than Jacobi's thought that immediate knowledge is philosophical knowledge which is to be counterposed to reason (*Vernunft*); and then they talk about reason and philosophy etc. like a blind man does about colours.' *Geschichte der Philosophie*, p. 544. The idea that religious faith is of more importance than rational knowledge of the content of that faith is said to be 'the most general standpoint of our time'. p. 545. For Hegel an important consequence of Kant's subordination of reason to the understanding is its powerlessness against the irrationalist philosophy of faith which makes only the empirically real

the knowable. As Hegel writes in *Faith and Knowledge*, '*Jacobi's philosophy* shares with Kant's the common ground of absolute finitude, both in its ideal form, as formal knowledge, and in its real form as an absolute empiricism. They also agree about the integration of these two absolute finitudes by way of a faith that posits an absolute beyond.' Hegel adds that within this common sphere, Jacobi and Kant form an antithesis: Jacobi's philosophy is a complete subjectivism, while in Kant's philosophy 'finitude and subjectivity have an objective form.' p. 97.
14. *The Phenomenology of Spirit*, p.10.
15. Thus Hegel writes: 'But philosophy is not meant to be a narration of happenings, but a cognition of what is *true* in them, and further, on the basis of this cognition, to *comprehend* that which, in the narrative, appears as mere happening.' *Science of Logic*, p. 588.
16. *Science of Logic*, pp. 204-212.
17. *Ibid.*, pp. 216-217.
18. Unlike Kant, Hegel has not figured in subsequent discussions in philosophy of mathematics. There is a very general article by Reinhold Baier, 'Hegel und die Mathematik' in *II Verhandlungen des Zweiten Hegelkongresses*, ed. B. Wigersma, (Tübingen: J. C .B. Mohr, 1932). There is also a useful discussion in Terry Pinkard's *Hegel's Dialectic: The Explanation of Possibility*, (Philadelphia: Temple Uni. Press,1988), pp. 41-54
19. *Logic*, para. 99. Nevertheless, Hegel stresses that measure as the synthesis of quality and quantity is the completion of the sphere of 'being', *Logic*, para. 107.
20. *Logic*, para. 48. Note that this definition follows directly from the epistemology of absolute idealism. This also runs parallel with the elimination of any need for a 'schematization', as Kant required, to legitimate metaphysical categories.
21. In theological language, the system is the description of the eternally active God, not a God that is beyond, but one that is immanent. But Hegel's God is ultimately the self-postulating *Logos*. In keeping with this is Hegel's substitution of a logic for traditional metaphysics and ontology. 'Accordingly, logic is to be understood as the system of pure reason, as the realm of pure thought. *This realm is truth as it is without veil and in its own absolute nature.* It can therefore be said that this content is the *exposition of God as He is in His eternal essence before the creation of nature and a finite mind'*, p. 50.
22. M.J. Petry calls Hegel's *Naturphilosophie* 'a sensitively structuralized, deeply informed and infinitely rewarding assessment of the whole range of early nineteenth century science.' 'Introduction' to *Hegel's Philosophy of Nature*, Vol.

1, (London: George Allen and Unwin, 1970), p. 60. Petry's notes certainly establish the breadth of Hegel's reading in the natural sciences. Milic Capek, on the other hand, writes that 'the tragicomedy of Hegel's philosophy of nature is that it was far *behind* the science of his own time.' 'Hegel and the Organic View of Nature', p. 109, in *Hegel and the Sciences*, ed Robert Cohen and Max Wartofsky, (Dordrecht: D. Reidel, 1984), *Boston Studies in the Philosophy of Science*, Vol. 64. According to Petry, Hegel's 'actual *mistakes* were few and far between.' *Op.cit.*, pp. 49. For the errors mentioned by Petry, see pp. 49-51. Capek, on the other hand, finds it incredible that Hegel could have spent 25 years studying Kepler and Newton and still made such errors of judgment. Amongst other things Capek criticises Hegel's claim that the law of gravitation can be derived from Kepler's third law. Hegel's 'proof' says Capek is 'based on an elementary confusion of symbols and the failure to understand the very meaning of mathematical proof', *op.cit.*, p. 110, and p. 119, footnote 4. From another quarter is Henry Paolucci's 'Hegel and the Celestial Mechanics of Newton and Einstein' in *Hegel and the Sciences*. Hegel is, for Paolucci, not only a profound philosopher of science, he is a precursor of Einstein, Planck and Bohr. Such wild claims, on the basis of very obscure passages and analogies, make Hegel something of a seer. Less exaggerated is David Lamb's *Hegel: From Foundation to System,* (The Hague: Martinus Nijhof, 1980.) In chapter 8 'Observation of organic nature: Inner as Outer and Outer as Inner', Lamb concentrates on similarities between Hegel and contemporary system's approaches to classification of organic material. It is, of course, understandable that Hegel's strength lies in its system's approach, as indeed did Schelling's. This is also Capek's conclusion.
23. *Encyclopaedia*, para. 246.
24. The Aristotelian roots of a *Naturphilosophie* are visible in Hegel's hierarchical and teleological description of nature, and the subordination of mathematical laws to the conceptual (i.e. 'logical') explication. Hegel's rejection of the idea of evolution is also indicative of an Aristotelian preference for 'rounded', rational forms. (A major difference between Hegel and Aristotle is Hegel's emphasis upon the continuity of the sciences.) For the rejection of evolution on the ground of its irrationality, see *Encyclopaedia*, para. 249. Kant had also considered evolution. But while Hegel dismissed it because it violated reason, Kant says only experience can decide if evolution is true, and that there is no experimental evidence to confirm it. *K.d.U.*, para. 80. Of course, Hegel is not advocating a retreat from the procedures of empirical science as a whole.

And to the extent that Hegel makes the framework of a science play a determinant part in scientific inquiry, Hegel, as a number of commentators have pointed out, is strikingly modern. See Gerd Buchdahl, 'Hegel's Philosophy of Nature and the Structure of Science' in *Hegel*, ed. Michael Inwood, (Oxford: University Press, 1985). Also by Buchdahl, 'Conceptual Analysis and Scientific Theory in Hegel's Philosophy of Nature (with Special reference to Hegel's Optics)' in *Hegel and the Sciences*. Petry's excellent, albeit overtly defensive, introduction to his translation of the *Naturphilosophie* also dwells on this (see the section 'Levels, Hierarchies, Spheres'), as does Lamb *op.cit.*, pp. 104-108. But modern discussions of theory-dependency forego any attempt to demonstrate a hierarchical necessity of the principles of the whole field of natural science. Science may be a unified field, but because of the complexity of the empirical sciences, the articulation of that unity is, in Kant's language, 'a mere idea', i.e. it would be an unending task to articulate that unity.
25. This is the sub-title of the Introduction.
26. Petry seems oblivious to the implications behind the claim that Hegel's *Encyclopaedia* is 'in a sense dualistic, since a distinction has to be drawn between its subject matter and its structure', *op.cit.*, p. 48. The dualism of structure and subject matter contradicts the Hegelian insistence upon the 'unity of form and content'. Throughout his commentary Petry also frequently suggests alternative developments between levels, again unaware of the ontological implications. It may make more sense to be a dualist of sorts, but, as we have argued throughout, Hegel's idealism takes its shape by its stand against dualism.
27. *What is Living and What is Dead in Hegel*, tr. Douglas Ainslie, (London: Macmillan, 1915 [1906]), pp. 169-170.
28. *Encyclopaedia*, para. 314.
29. See Petry, *op. cit.*, p. 81. The most famous scientists normally numbered within the Schellingian circle are Oken, Ritter and Orsted.
30. Dilthey, *Gesammelte Schriften*, Vol. IV, p. 253.

4 The substantiation of freedom: a metaphysics of the state

From its very beginning, Hegel's metaphysical dispute with Kant revolved around the idea of freedom. As a theological student Hegel had emphasized the superiority of Jesus's ethical teaching to Kant's because of its reconciliation between nature and duty in the concept of love.[1] He had also emphasized that the Enlightenment conception of ethics had not provided a cultural substitute for religion. Religion, said the young Hegel, was not merely a 'science' of what was rational in the concepts of God and the soul.[2] In a religious community (*Gemeinde*) duties were not prescriptions of a mind estranged from one's social activity, but the duties were inseparable from the feelings and behaviour of its members.

The critique of Kant and Enlightenment expressed in Hegel's early theological works emphasized the idea of freedom that Hegel was to hold all of his life: the idea of freedom must be substantial, grounded in the community (itself the expression of spirit) and not merely a moral idea which has as its ideal a standard beyond the world.

In a number of early publications Hegel attacked what he believed to be the one-sided conception of freedom accompanying the dualist separation of nature (necessity) and reason (freedom) in the philosophies of Rousseau, Kant, and Fichte. The conception of moral freedom as the absolute autonomy of a will which transcended the irrational world of nature was, observed Hegel, determined by the very mechanistic philosophies it had sought to transcend. In response to a world bereft of spirit/reason, the natural right theories of Rousseau, Kant and Fichte had constructed an idea of spirit/ reason which had no point of contact with the world. The attempt, for Hegel, to make the world conform to an 'idea' of this one-sided conception of reason was necessarily oppressive. In *Faith and Knowledge* Hegel spoke of Kant's dualism as:

> the empty concept in its unmoved opposition to nature, [which] can produce nothing but a system of tyranny [of the law of reason over human nature], and a rending of ethical life (*Sittlichkeit*) and beauty, or, else like Kantian morality cleave to so-called duties of a formalistic kind that determine nothing.[3]

In his previous work, *The Difference Between the Fichtean and Schellingian Systems of Philosophy*, Hegel had argued that Kant's theory of natural right inexorably led to Fichte's oppressive utopia.[4] Hegel pointed to the social oppression and social discord resulting from the picture of the state as a lifeless machine, the product of an autonomous and indeterminant will.

> Through the absolute opposition of the pure and the natural drive, natural right becomes a presentation of the total dominance of the understanding and of the slavery of the living. It is an edifice in which reason plays no part and which it thus rejects because it must exist most explicitly in the most perfect organization which it can give itself, in the self-formation into a people (*einem Volk*). But that state based upon the understanding (*Verstandesstaat*) is not an organization but a machine. The "people" is not here the organic body of a common and rich life, but an atomistic, life-impoverished multitude. Its elements are absolutely opposed substances, in part a set of points, the rational beings, and in part materials variously modifiable by reason. That is, in this form, through the understanding, they are elements the unity of which is a concept, the connection of which is an endless domination.[5]

And in the the *Phenomenology* he drew the connection which Robespierre himself had drawn (in his address to the French Convention of Feb. 5, 1794) that the Terror was 'an emanation of virtue'. For Hegel, the virtue of a will too good for this world was a virtue that could only decimate the earthly forms of the ethical life.[6]

As the above citation from the *Difference* indicates, Hegel shared the organic view of the state and that was being espoused in the state theories of Schleiermacher, Novalis, Schlegel, and Adam Müller that Burke had advanced against defenders of the 'Rights of Man' His reasoning for this advocacy was similar to his contemporaries and, again, Burke: the content of people's wills derives from the very ethical spheres which give people their identity.[7] These spheres are maintained because codes of behaviour are adhered to, and each role is regulated by the roles of others. The role itself determines what is right and wrong; what, for example, is a good husband/father/child etc. For Hegel, any attempt to rely upon the conscience or a formal law such as Kant's moral imperative is potentially destructive. Instead of compliance with a definite duty that must be performed if an ethical relationship is to be maintained, the individual is given too much scope in providing a content for a universal law. For Hegel, the appropriation of Kant's moral theory by Fichte, then Fichte's by Schlegel demonstrated how easily Kant's moral formalism was transformed into a defence of the immoral, as was evident in the threat that Schlegel's had posed to the stability of the family by dismissing the necessity of the marriage ceremony. (It was, for Schlegel, a mere externality of the state.[8])

In spite of Hegel's critique of the one-sidedness of the post-Rousseauian dualist metaphysics of natural right, Hegel emphasized that the natural right theorists had correctly seen that the concept of freedom is associated with the concept of will, and that the *logical* grounding of the concept of right must be in the free will.[9] But Hegel saw it as disastrous to separate the prescripitions of the will (practical reason) from the substantial ethical relations (of the family, civil society, and the state). A philosophical account of right had to specify the unity between the abstract and moral basis of right and the substantive ethical relations.

In so far as the actual forms of the ethical life were to provide the real end of right, Hegel emphasized that his position was not to be confused with the positivist concept of right which collapsed the distinction between the substantial and accidental, thus equating might and right.[10] What needed to be demonstrated was that right was not something intrinsically unattainable, an ideal always too good for this world, rather the 'idea' was here and now. Conversely, the world was not 'God-forsaken', a place bereft of justice. Justice was present in the world. Philosophy had to affirm and delineate the shape of its presence.[11]

For Hegel, this meant specifying the necessity of the connections between various moments of the concept of right. The first of these ('abstract right') affirmed the act of possession and the contractual relations requiring recognition of the person as a willing agent. The second, that of morality (*Moralität*), affirmed the principle of moral intention in which the particular subject recognizes as moral what it deems to be universal. The third and final moment, the idea of freedom 'as the living good', is that of *Sittlichkeit*, the ethical life.[12]

For Hegel, the original (logical) relationship in which the will receives a determinant ethical content is in the family. Underpining this relationship is, what Hegel considers to be the 'immediate' ethical feeling, love. The immediate feeling of romantic love leads to marriage and a sense of shared identity, mutual responsibility and the necessary allocation of roles in order for the relationship to be maintained. Hegel 'deduces' that the way the relationship is to be maintained is through monogamy and patriarchy. No particular person determines the essential nature of the family, although the stability of the family depends upon the behaviour of its individual members. Rather the wills of the particular persons are determined by the relations which constitute a family. As Hegel says the individuals are accidental.[13]

The second shape of ethical life is civil society.[14] In this sphere individuals are primarily constructed as systems of wants, economic agents seeking to satiate and develop their personal inclinations and seeking to sell their individual skills, i.e. their capacities to satisfy the wants of others. This sphere is characterized by its schismatic nature. But, unlike Marx, Hegel holds that civil society is intrinsically rational. He has two basic reasons for this.

Firstly, it is only within the context of civil society that *man* as a concrete representation, defined by rights and duties, comes into existence.

> In [abstract] right, what we had before us was the person; in the sphere of morality, the subject; in the family, the family member; in civil society as a whole, the burgher or *bourgeois*. Here at the stand-point of needs...what we have before us is the composite idea which we call *man* (*Mensch*). Thus this is the first time, and indeed properly the only time, to speak of *man* in this sense.[15]

Secondly, the multiplying effect of the needs of civil society plays a special role in the widespread development of consciousness. It is thus indispensable for the education/cultivation (*Bildung*) of mind, which is the real end of Hegel's enterprise. The way in which mind is developed in civil society is the necessary consequence of the transformation of nature that takes place through work. Nature loses its strict identity as it becomes subordinate to the needs that are developed in civil society and the techniques of production. Nature is the raw material upon which mind sets to work for the creation and satiation of social needs.[16] For this reason Hegel emphasizes the dual relationship between theoretical and practical education (*Bildung*). Labour not only requires the existence of the materials to be reformed. It also needs thought. It is only through thought that the rules and procedures of transforming natural object A to worked-on object B are carried out by the labourer. That these procedures become 'second nature' illustrates for Hegel that the workers, unbeknown to themselves, become vehicles in the self expansion or cultivation (*Bildung*) of spirit.[17]

The rationality of civil society for Hegel is thus apparent insofar as: (1) reason develops through it, and (2) individuals develop their capacities and market them. Yet in spite of the rationality of civil society, Hegel, as several commentators of Hegel's political philosophy have observed, saw civil society as threatening the very existence of communal life. Civil society brings with it the unrestricted creation and expansion of social wants and the division and mechanization of labour. For Hegel, this division of labour brings with it two social disorders: it dehumanizes the labourers, and it inevitably leads to economic crises.[18] This occurs, for Hegel as it also did for Marx, partly through labour displacement by machinery, and partly through over-production. Once thrown into poverty, Hegel fears that the industrial labourers may become a mob (*Pöbel*),[19] destroying the social harmony of ethical life. In other words civil society nurtures within it the necessity of poverty and the possibility of future days of Terror. Thus, for Hegel, the question of the abolition of poverty 'is one of the most disturbing problems which agitate modern society.'[20]

For Hegel, this and other problems arising from the private pursuit of (economic) liberty can only be resolved within a higher more complex sphere of the ethical life, the state. The state protects family life and civil society. It also develops civic virtues, mainly through the education system and (in times of war) military service. Also crucial to Hegel's understanding of the state is his judgment that 'the people' is a meaningless abstraction unless it refers to the specific powers/attributes that accrue to it through its national identity and the institutional forms of the sovereign, the executive and the legislature, i.e. the institutions which provide a unity of social articulation, and a body for articulating, debating and enforcing the legal boundaries and bindings of the community.

A point that has been made in various ways by Hegel scholars, and which warrants pausing upon, is the importance that the business class play in Hegel's state. Indeed, Hegel creates the impression that he is building a theory of the modern state around the business class, as if the major problem of the modern state was specifying the function and constraining the power of the business class.

The proliferation of the division of labour and the increasing quantity and variety of material goods that resulted from that division was intrinsically connected to the expansion of private property and hence to the increased power of the bourgeoisie. The French revolution demonstrated to Hegel that the economic power achieved by this class would inevitably be translated into political power. Likewise, the revolution had made it clear that any attempt either to deny political power to the bourgeoisie or to suppress the public expression of its ideas (through a free press) was bound to fail.[21] Hence for Hegel the fate of the modern state was intrinsically connected to the aspirations and power of the bourgeoisie.

The issue, then, was not whether the bourgeoisie should hold political power, but the identification of the adequate form it should take. Because ethical motivations of the business class were primarily based on self-interest and commercial gain, Hegel thought that neither the sovereign nor the executives of the modern state could come from its ranks. When this is combined with Hegel's convictions that only an institutional form provides ethical content and that the restoration was the unavoidable consequence of the revolution, Hegel had little choice but to locate the source of sovereignty where it had traditionally been located, in the monarch.[22] At the same time, Hegel was equally convinced that monarchical absolutism was a spent force, irreconcileable with the modern spirit of freedom. Hegel's compromise of a constitutional, yet hereditary monarch was a neat piece of reconciliation, yet as unsatisfactory for conservatives, as it was for liberals and democrats.[23]

The power of the business class was also to be controlled by a distribution of power to those classes whose interests were not locked into 'the system of needs.' For Hegel there are two 'classes' that possess this independence: the 'class' of civil servants and the 'class' of landed property owners.[24] The executive (the civil servants) and the members of the upper chamber of the legislature were seen by Hegel as 'natural allies' against the self-interested class.

The stabilising element of business life was, for Hegel, not only extrinsic to the business class, but it was necessary to provide internal stability for civil society. For Hegel, the corporation partly provided this stability.[25] The function of the corporation is largely educational in so far as it trains its members in the relevant skills necessary for their occupations. It also provides its members with a sense of identity that is greater than the self-interest of the particular member. It is a stepping stone to a greater sense of awareness, the awareness that one has duties not only to oneself, one's family and one's peers, but to the community as a whole. These duties as well as individual rights are made actual in the laws of the state.

The corporation also provides financial support for those members who experience economic hardships.[26] Because of the common interest that binds members of a corporation Hegel argues that if the legislature is to be rational, then the legislature must be composed of representatives of corporations, and not merely political representatives who have no real identity with the interests of their electors.[27]

Excluded from the corporations and thus from political representation are the unskilled industrial workers and the day-labourers. That is to say the potentially revolutionary class, the class which becomes a 'rabble' in times of crisis, cannot be helped by civil society because it is through the inviolable rules governing civil society that the social crisis and the rabble are created. Hegel considers a number of options which could alleviate the revolutionary threat such as direct taxation of the wealthier classes, or making the wealthier classes fund institutions to help those in need.[28] But he sees these options as directly contradicting the principle governing civil society. The poor would be provided for without having to work. This in turn, claims Hegel, would lead to the lack of independence, individuality and honour, i.e. essential sentiments of civil society.[29] He also considers the possibility of the state providing work. But, Hegel points out, this can only aggravate the problem. For this would increase the volume of production. And it is the superfluity of wealth of civil society, over-production, that is responsible for the misery in the first place.[30]

The solution that Hegel sees as the one in keeping with the rules of civil society is the expansion of markets and colonization.[31]

A closer consideration of the issues involved here indicates that Hegel not only sees the modern state as necessarily responsive to the internal schisms of civil society, the modern state is unavoidably forced into conflicts with other states in the same predicament. War, for Hegel, is ultimately an indispensable 'moment' of the concept of the modern state. The *Philosophy of Right* presents war as a rational necessity for the substantiation of freedom.

From his earliest writings on politics, Hegel held the belief espoused by Machiavelli, and even articulated by Kant, that peace and prosperity bring with them a decay in ethical life. And like Machiavelli he believed that war shakes up a people, instilling virtues of bravery and providing a higher purpose as the bonds of community are reforged out of necessity. In the *Philosophy of Right* Hegel says that peace makes the people see themselves only as particulars, and as particulars their substantive freedom is endangered.[32]

For Hegel war engages all members of the state in a common ethical purpose. If war subordinates all of particularity and finitude to the totality of communal life (the state) it is, for

Hegel, absurd to condemn it. The substance of freedom is not to be determined by a mere ought, a mere moral freedom.33 Moreover, for Hegel, war not only strengthens the character of a people, it provides the means to avoid the possibility of revolution: 'As a result of war, nations are strengthened, but peoples involved in civil strife also acquire peace at home through making wars abroad.'34

In keeping with this position, unlike Kant, who observed and condemned the conquests and brutalities of the Europeans against native peoples, Hegel accepts such conquests as essential to the realization of freedom.35 The conquered people, the barbarians, have no substantive rights.36 They are caught up in 'the struggle for recognition'. They are passing moments on the stage of world history. In light of the state being the ethical ground of a people, there is no way for Hegel that it is meaningful to condemn the conquest of other peoples.

In summing up Hegel's conception of a rational unity we may say that unity may only be preserved by war. Concomitantly real freedom can, for Hegel, only be preserved by a state which is a moment within the greater dynamic of world history. Not only is the state the embodiment of freedom, but history itself is the vehicle of freedom.

Kant had already moved towards a teleological conception of history in his idea of the 'civic constitution' and political freedom as part of the 'hidden purpose' in history.37 But when Kant had raised this problem, it was proposed within a dualist framework. Kant's view of history as teleological was a regulative idea. But for Hegel, the idea of a teleological view of history could not be, as it was in Kant, merely an idea for arousing a sense of optimism and a feeling of awe for our duties and the world in which they are to be performed. Having eliminated the cleavage between regulative and constitutive principles, Hegel had to affirm war and colonization (albeit not every war, and not every barbarism perpetrated in the colonies) if he was to remain true to his metaphysic. The state and history were the embodiments of reason, the expressions of 'the divine will'.38

Notes

1. 'The Spirit of Christianity and Its Fate' in G.W.F. Hegel, *Early Theological Writings*, tr. T.M. Knox, intro. Richard Kroner, (Philadelphia: Uni. of Pennsylvania Press, 1971), pp. 205-224.
2. See 'The Tübingen Essay (1793)' in G.W. F. Hegel, *Three Essays 1793-1795*, tr. Peter Fuss and John Dobbins, (Notre Dame: Uni. of Notre Dame Press, 1984), p. 32, also see pp. 39-49. The social dimension of religion for Hegel is also

clearly visible in 'The Positivity of the Christian Religion' in *Early Theological Writings*, see esp. sect. 21. For a fuller account of the social and political dimensions of Hegel's theological writings see ch. 1 of Shlomo Avineri's *Hegel's Theory of the State*, (Cambridge: University Press, 1972). Lukács' *The Young Hegel* also stresses the politics of Hegel's theological writings. Lukács's work was partly undertaken to remedy what he believed were distortions in Hegel scholarship. In particular it sets itself against Dilthey's and Kroner's readings. While it succeeds in emphasizing a region insufficiently addressed by either Dilthey or Kroner, it should be said that it is particularly unfair to Kroner. Kroner's massive work is summed up in Kroner's sentence torn from context, that 'Hegel is the greatest irrationalist known to the history of philosophy.' (*Ibid.*, p. 400, see also xviii.) For the context see Kroner, *Von Kant bis Hegel*, Vol. 2, pp. 267-272.
3. *Faith and Knowledge*, p. 143.
4. G.W.F. Hegel, *Natural Law: The Scientific Ways of Treating Natural Law, Its Place in Moral Philosophy, and Its Relation to the Positive Sciences of Law*, tr. T.M. Knox, (Philadelphia: Uni. of Pennsylvania Press, 1975), p. 124.
5. *Ibid.*, pp. 65-66.
6. See para. 5 of the *Philosophy of Right*. This paragraph needs to be considered with the chapter 'Absolute Freedom and Terror' in the *Phenomenology*.
7. On the importance of Burke in Germany see Ch. VIII of Reinhold Aris's *Political Thought in Germany: From 1789 to 1815*, (London: Frank Cass, 1965 [1936]). See esp. p. 257. On the relationship between the organic theory of the state and the conservative intellectual milieu in Germany (a milieu in which one time supporters of the revolution such as Schlegel conspicuously figured) see p. 294. The organic conception of the state is central to Schleiermacher, Adam Müller, Novalis and Schlegel, as well as Hegel. See Aris, ch. X. Useful selections of the political writings of these thinkers can be found in H. S. Reiss's *The Political Thought of the German Romantics 1793-1815*, (Oxford: Basil Blackwell, 1955). Another valuable collection of the ideas of the period is *Gesellschaft und Staat im Spiegel deutscher Romantik*, ed. Jacob Baxa, (Jena: Gustav Fischer, 1924). Although Hegel vigorously attacked Schlegel, Schleiermacher, and other German romantics, there was a common bond between Hegel and these thinkers. This is evident in their shared vocabulary and their shared intellectual priorities gravitating around the core idea of the world as revealing the presence of spirit and the conceptual importance of *Bildung*. All these thinkers emphasized the importance of the family, religion and the state as the

substantive basis of ethical behaviour. As Hegel disclosed in a letter to Franz Baader, he believed that he and Baader were united in their fundamental preoccupations with spirit, and these common preoccupations were far more important than their differences. See Hans Grasl, 'Hegel an Baader', *Hegelstudium*, 1963 (Bonn: Bouvier, 1963), Vol. 2. Nevertheless, there were irreconcilable differences over the status of faith. For Hegel this was merely the retention of a beyond (*Jenseits*). To both Baader and Schlegel, Hegel's subordination of faith to reason, and religion to the state made Hegel an atheist. Schlegel even calls the 'unfolding' of Hegel's *Weltgeist* 'the development of the Anti-Christ', see Friedrich Schlegel, *Kritische Ausgabe*, ed. Ernst Behler (München: Paderborn, 1975), Vol. 22, p. 69, also Vol. 30, (1980), pp. 320-321. The importance of 'faith' to Schlegel even enters into the his conception of the state. 'The state', he says, 'is based on faith.' *Op.cit.*, Baxa, p. 74. For further discussion of Schlegel and Hegel see Ernst Behler, 'Friedrich Schlegel und Hegel' in *Hegel-Studium*, Vol. 2.

8. *Philosophy of Right*, para. 140. In the section of the *Phenomenology* entitled 'Spirit that is certain of itself. Morality', Hegel explores the dialectical development of consciousness from the formal moral position of Kant through to the 'moral' aestheticism which makes play with the substantive content of good and evil. In 'Die Beisetzung der Romantik in Hegels *Phänomenologie*' E. Hirsch identifies the philosophical and literary defenders of the (im)moral positions described by Hegel. See *Deutscher Vierteljahrschrift f. Literaturwissenschaft u. Geistesgeschichte*, 2, 1924, pp. 510-524. Also see *Philosophy of Right*, para. 140, and, on Schlegel's concept of irony, see Hegel's *Aesthetics: Lectures on Fine Arts*, Vol. 1, tr. T.M. Knox, (Oxford: Clarendon,1975), pp. 64-69. For the critique of *Lucinde* and Schleiermacher's defence see *Philosophy of Right*, para. 164 and the Addition. Hegel, however, is closer to Schlegel than is at first apparent. They are united by the ethical importance they ascribe to the family and the forms of ethical life as opposed to the moral *a priori* reasonings of natural right theory. In para. 270 of the *Philosophy of Right*, Hegel attacks the position held by Schlegel, that the state should be subordinate to the church. Yet Hegel requires that all citizens be members of a church. To this extent he shares Schlegel's belief that religion is an indispensable component for the formation of ethical life. I would like to thank Tom Morton, for his helpful discussions with me on the German romantics.

9. Hegel's affinity with natural law is evident in the alternative title of *Philosophy of Right* — *Natural Law and Political Science in Outline*.

10. See esp. the lengthy critique of Haller, *Philosophy of Right*, para. 258. Also *Natural Law: The Scientific Way of Treating Natural Law*, pp. 130-131. In spite of Hegel's criticism here we should not overlook the affinity Hegel shares with the historical school of right: the belief that the law is the expression of the consciousness of the people (*Volk*). The major point Hegel holds against the historical school of law is the belief expressed by Savigny in the undesirability of codifying the law. By insisting upon the codification of the law, Hegel believes it is possible to avoid the equation of the contingent, the brute positive, with the authentically (i.e. rational) positive. One is also led to interpret the 'people' only as concretely defined by their institutions. The law is not subject to drastic changes on the basis of momentary community feelings. This is not only relevant to Hegel's difference with the historical law school, but to his anti-atomistic critique of liberal democracy. Also on Hegel and the historical law school see Norbert Bobbio's excellent essay 'Hegel und die Naturrechtslehre' in *Materielien zu Hegels Rechtsphilosophie*, Vol. 2, ed. Manfred Riedel, (Frankfurt/M: Suhrkamp, 1975), pp. 85-86.
11. Those who see the world as 'God-forsaken' rely upon their own (for Hegel, irrational) feelings to specify how the state should be. *Philosophy of Right*, pp. 3-6,12. For Hegel it is not the task of philosophy to prescribe to the state what it should be: 'the instruction which it [philosophy] may contain cannot consist in teaching the state what it ought to be; it can only show how the state, the ethical universe, is to be understood.' p. 11.
12. *Philosophy of Right*, para. 142.
13. *Philosophy of Right*, para. 145.
14. For a discussion of the historical antecedents and transformations of the concept of 'civil society' see Manfred Riedel, '"State" and "Civil Society": Linguistic Context and Historical Origin' in *Between Tradition and Revolution: The Hegelian Transformation of Political Philosophy*, tr. Walter Wright, (Cambridge: Uni. Press,1984).
15. *Philosophy of Right*, para. 190. It should be noted here that Hegel is specifically endorsing the idea later held by Marx that the theory of modern natural right is governed by a specific social form, bourgeois society. There is, for Hegel, a necessary relationship between the social relations of a period and the ideas of the period. However, unlike Marx, Hegel is attempting to specify the relationship between the different levels of the concept of 'right/law'. He finds the elementary ground of the concept in the will, because only a will can designate something as right or wrong. The content of the historical and social conditions in which it

operates are ultimately, for Hegel, inseparable from reason and thus what the will is.

16. 'There is', says Hegel, 'hardly any raw material which does not need to be worked on before use. Even air has to be worked for because we have to warm it. Water is perhaps the only exception, because we can drink it as we find it.' *Philosophy of Right*, Addition, para. 196. The idea of the plasticity of nature, and its subordination to labour was fundamental to classical political economy. That subordination is clearly expressed by Locke when he says 'I think it will be but a very modest Computation to say, that of the *Products* of the Earth useful to the Life of Man 9/10 are the *effects of labour.*' *Two Treatises of Government*, 'The Second Treatise', para. 40. The importance of Hegel's relationship to political economists in the *Philosophy of Right* is common knowledge. It is emphasized in interpretations by Lukács, *op.cit.*, Raymond Plant, *Hegel*, (London: George Allen and Unwin, 1973), and Bernard Cullen, *Hegel's Social and Political Thought*, (Dublin: Gill and Macmillan, 1979). Also see *Hegel on Economics and Freedom*, ed. William Maker, (Macon: Mercer Uni. Press, 1987), Norbert Waszek, *The Scottish Enlightenment and Hegel's Account of 'Civil Society'*, (Dordrecht: Kluwer, 1988), Lawrence Dickey, *Hegel: Religion, Economics and the Politics of Spirit,* (Cambridge: Uni. Press, 1987). It is not often appreciated that Hegel's preliminary definition of nature in the *Encyclopaedia* as the 'form of otherness (*Anderssein*)' stands in the closest relationship to the idea of nature as the material of practical (and hence social) transformation. Thus not only does the *Logic* close with the 'the absolute Idea' in which theoretical reason (knowledge) and practical reason (will) are reconciled as self-realized reason, but the *Philosophy of Nature* commences with the unity of practical and theoretical reason. In para. 245 of the *Encylopaedia* Hegel writes: 'In the practical relationship which man establishes between himself and nature, he treats it as something immediate and external; he is himself an immediately external, and therefore sensuous individual, who is nevertheless also justified in acting as purpose in the face of natural situations.' *Philosophy of Nature*, Vol. 1. For the most part, civil society provides the ground for the specific determinations of the will imprinting itself on raw nature, and thus for the material products possessing value in society.

17. See *Philosophy of Right*, para. 197. In this integration of labour and intelligence, we see once again how theoretical reason takes on a social significance for Hegel that lies completely beyond the scope of Kant's questioning. For Hegel, the cultivation of theoretical reason (Hegel says 'the

education of the understanding in every way, and so also the building up of language') is accelerated within the sphere of civil society.
18. *Philosophy of Right*, para. 198. Hegel had first systematically expressed his ideas, based upon his readings of political economy, while at Jena in the unpublished *Jenaer Realphilosophie*, *Hegel and the Human Spirit: A Translation of the Jena Lectures on the Philosophy of Spirit 1805-06*, tr. Leo Rauch, (Detroit: Wayne State Uni. pp. 139-140), and *Jensener Realphilosophie*, in Gerhard Göhler's collection *Georg Wilhelm Friedrich Hegel: Frühe politische Systeme*, (Frankfurt/M: Ullstein, 1974), pp. 236-240. Page numbers provided are to the more popular editions by Johannes Hoffmeister, *Sämtliche Werke*, Vol. 19, 20, (Leipzig, 1931,1932). Göhler's collection also includes the *System der Sittlichkeit* as well as his excellent commentary and selections and essays by Rosenzweig, Marcuse, Lukács, Ilting, Habermas and Riedel. Since the discovery of these lectures, the similarity of Hegel's description of the degradation of the factory worker has frequently been compared with Marx's conception of alienation. Had Marx read these manuscripts he could not have upheld his view that Hegel conceives only the positive side of labour, and that Hegel knows only abstract mental labour. Karl Marx, *Early Writings*, tr. Rodney Livingstone and Gregor Benton, (Harmondsworth: Penguin, 1975), p. 386. This claim by Marx also does not square with Hegel's conception of crisis in the *Philosophy of Right*.
19. Poverty is not a sufficient condition for the creation of a mob, there must also be the accompanying feeling of indignation. *Philosophy of Right*, para. 244 and Addition.
20. *Philosophy of Right*, Addition, para. 244.
21. In *The Philosophy of History*, Hegel makes it clear: (1) that he defends the principles of the revolution, and (2) that the revolution was an act of necessity brought about by the degeneracy and intransigence of the privileged estates, pp. 446-453. For Hegel tradition was not acceptable merely because it was tradition. In the 'Proceeding of the Estates Assembly in the Kingdom of Würtemberg, 1815-1816', Hegel attacks the landed estates of Würtemberg for having 'slept through' 'the richest period of world history that the world has probably ever had.' He reminds them: '"Old rights" and "old constitution" are such fine grand words that it sounds impious [to contemplate] robbing a people of its rights. But age has nothing to do with what "old rights" and "constitution" mean or with whether they are good or bad. Even the abolition of human sacrifice, slavery, feudal despotism, and countless [other] infamies was in every case the cancellation of something that was an "old right."'

Hegel's Political Writings, (Oxford: Clarendon,1964), tr. T.M. Knox, intro Z.A. Pelczynski. Hegel's endorsement of the French revolution is covered in detail by Joachim Ritter in *Hegel and the French Revolution: Essays on the Philosophy of Right*, tr. Richard Dien, (Cambridge Mass.: MIT Press,1982). However Ritter tends to downplay Hegel's conservatism.

22. For the inevitability of a restoration accompanying revolution, see *The Philosophy of History*, p. 453.
23. The fact that Hegel had endorsed constitutional monarchy rather than absolute monarchy was seen by some Prussian patriots as indicative of Hegel's subversive teaching. See K. E. Schubarth's 'Über die Unvereinbarkeit der Hegelschen Staatslehre mit dem obersten Lebens- und Entwicklungsprinzip des Preussischen Staats' written in 1839, *Materialien zu Hegels Rechtsphilosophie*, ed. Manfred Riedel, (Frankfurt/M:Suhrkamp, 1975), pp. 249-65. To conservatives Hegel was too liberal, but to many liberals Hegel was too conservative. The most famous liberal critique of Hegel came from Rudolph Haym, his critique of the *Philosophy of Right* from *Hegel und Seine Zeit*, and Rosenkranz's classic defense are also included in *Materialien*, Vol. 1. A common critique of the *Philosophy of Right* is that Hegel is merely an apologist for the Prussian state. To this criticism, Eric Weil has replied that the Prussian state of Hegel's time was one of the most progressive states in Europe. See his *Hegel et L'Etat*, (Paris: Vrin, 1950).
24. For Hegel, the three basic classes are the agricultural, business and universal class (i.e. the civil servants). These three classes supposedly correspond to the moments of knowing. The agricultural class embodies the immediacy of knowing; the business class corresponds to the ways of thinking embodied in the metaphysics of the understanding (*Verstandmetaphysik*). They are conceived as a reflecting and formal class and the civil service is classified as the universal class. *Philosophy of Right*, para. 202. For Hegel, this division into forms of social consciousness is indicative of the way in which the problem of politics is set as one of the conflicting forms of consciousness, and how consciousness in turn is structured by necessary (onto)logical categories. Needless to say, Hegel's connections here are mere analogies. That Hegel can lay such importance upon such analogies is indicative of the problem of the speculative method, or for that matter, any method which wants all parts to conform to a 'total' plan.
25. The corporation is called by Hegel 'the second ethical root of the state'. *Philosophy of Right*, para. 255. The institution of marriage is, for Hegel, the first ethical root. It is also 'a

bulwark' against the atomistic and divisive spirit of civil society.
26. *Philosophy of Right*, para. 252, 253.
27. *Philosophy of Right*, para. 311. As soon as one considers the specific details of Hegel's rational state an obvious problem emerges, a problem which was picked up by two of Hegel's contemporary reviewers of the *Philosophy of Right*, H. G. Paulus and J. F. Herbart. Hegel's rational state, which is not supposed to be in a beyond, which is not supposed to prescribe what the state should be, is not anywhere (although it has more in common with Prussia than most other states). Paulus reasonably asked 'Where, then, is *Rhodes*, where the philosopher is supposed to do his *political* dance — is it in Germany, or France, England or Spain?' He then adds that Switzerland, North America and Asia appear to be bereft of reason. *Materialien*, Vol. 1, pp. 63-64. Also see Herbart, *ibid*, p. 97. Scholars today usually interpret the institutions proposed by Hegel in a very plastic manner so that the *Philosophy of Right* retains its relevance for us. However, such a charitable interpretation of the *Philosophy of Right* easily leads to an *ad hoc* interpretation of Hegel.
28. *Philosophy of Right*, para. 245.
29. *Ibid.*
30. *Ibid.*
31. *Ibid*, para. 256-248.
32. *Philosophy of Right*, para. 324. For Kant see *K.d.U*, para 28. In para. 324 Hegel cites his earlier formulation from the *Natural Law: The Scientific Ways of Treating Natural Law*. In his chapter on Hegel in *Machiavellism: The Doctrine of Raison D'Etat and its Place in Modern History*, (New Haven: Yale Uni. Press, 1957), Friedrich Meinecke draws attention to Machiavelli's influence upon Hegel, which is particularly noticeable in Hegel's early work *The German Constitution*. The role of war in Hegel's system has suffered from the debate over whether Hegel was a pioneer of the totalitarian state. (Some examples are included and criticized in a collection of essays edited by Walter Kaufmann, *Hegel's Political Philosophy*, New York, 1970). This debate led to the majority of Hegel scholars adopting highly defensive (and excessively liberal) interpretations of Hegel's attitude to war to compensate for the overt hostility and often (but not always) ill-informed attacks by Hegel's critics. The defensive tone is evident in the chapter on war by Avineri, *Hegel's Theory of the Modern State*, and in E. H. Harris's 'Hegel's Theory of Sovereignty, International Relations and War' in *Hegel's Social and Political Thought: The Philosophy of Objective Spirit*, ed. Donald Verene, (New Jersey: Humanities Press, 1980). Also defensive, but from a quasi-

Marxian perspective, is Jacques D'Hondt's 'L'Appréciation de la guerre révolutionnaire par Hegel' in D' Hondt, *De Hegel à Marx*, (Paris: Presses Universitaires de France, 1972). Carl Friedrich's 'Introduction' to his selections of Hegel's writings, *The Philosophy of Hegel*, (New York: Random House, 1954) well demonstrates that those who saw totalitarian elements in Hegel were not all ignorant of the intricacies of Hegel's thought. The following account agrees largely with Paul Thomas's comments to the papers by Harris and Paolucci in *Hegel's Social and Political Thought*. In particular I agree with his remark that 'Hegel makes war *the* significant integrative institution of the modern state', p. 174. Also see Steven B. Smith, *Hegel's Critique of Liberalism*, (Chicago: Uni. Press, 1987), pp. 156-164.

33. The major philosophical exponent of the position Hegel is attacking, Kant, makes perpetual peace the ideal towards which all states should strive. For Hegel, Kant's position on war and the accompanying idea of a federation of nations is one further illustration of remaining by an empty 'ought'; it subordinates the substance of right to an abstraction. For Hegel's critique of Kant's idea of perpetual peace and the federation of states, see *Philosophy of Right*, Addition to para. 324, and para. 333. One should not forget that Kant also believed that culture develops through war and the fear of future war. Kant claims that perpetual peace was only possible in a perfect culture, which will exist 'God alone knows when'. 'Conjectural Beginning of Human History', in Immanuel Kant, *On History*, tr. Lewis White Beck, p. 121. Page numbers refers to marginal numbers which refer to the *Akademie* edition of Kant's works. For Kant, war is a natural condition and peace an artificial one. 'Perpetual Peace' in *On History*, pp. 348-349. Hegel's disagreement with Kant again comes back to Hegel's desire to reconcile reason and the world.

34. *Philosophy of Right*, Addition to para. 324. Hegel's conception of the rational necessity of war employs an economic premiss not shared by Kant and other defenders of the French Revolution: the spirit of commerce is antagonistic and not intrinsically peaceful. Commerce, for Hegel, leads to war, firstly within civil society itself, and then in the founding of colonies, and inevitably with other modern nation states. According to Kant 'the spirit of commerce...is incompatible with war.' *On History*, p. 368. Kant's assumption, common enough among liberal opponents of the aristocracy, would seem to be that wars are fought to maintain the sumptuous lifestyle of a non-productive class, the aristocracy. The same liberal sentiments were expressed by Thomas Paine who wrote: 'If

commerce were permitted to act to the universal extent it is capable, it would extirpate the system of war, and produce a revolution in the uncivilized state of governments. The invention of commerce...is the greatest approach towards universal civilisation, that has yet been made by any means not flowing immediately from moral principles.' *The Rights of Man* (Harmondsworth: Penguin, 1985), pp. 212-213.

35. For Kant's condemnation see 'Perpetual Peace' in *On History*, pp. 358-359.
36. *Philosophy of Right*, para. 351.
37. In the third *Critique*, 'Perpetual Peace' and 'Idea for a Universal History from a Cosmopolitan Point of View', Kant had claimed that war will eventually force people to erect constitutions which will lead to perpetual peace and a federation of nation states. This claim is part of Kant's 'critical' strategy of reconciling nature and reason, as nature is interpreted as purposeful and as contributing to the fulfilment of human freedom. The idea, as has been frequently noticed, foreshadows Hegel's concept of the cunning of reason. However, once again Kant's reconciliation between the purpose of nature and freedom is cast within a dualistic philosophical framework. For the purposefulness of war cannot, for Kant, be proven with certainty. Instead, the idea is an *a priori* principle for observing history, a regulative idea enabling us to read history as the unfolding of moral perfection. See Kant's essay, 'Idea for a Universal History from a Cosmopolitan Point of View', where Kant talks of using a 'guiding thread (*Leitfaden*)', pp. 17, 29-31. In spite of the sublimity and natural purpose of war, it remains, for Kant, a condition of nature to be eliminated.
38. In so far as philosophy is allied with universities, even philosophy, for Hegel, is indebted to the state. This is a major point in Hegel's bitter polemic against Fries. For Hegel, Fries had abused his public position by encouraging discord within the state. Hence for Hegel the state was fully within its rights to remove him from his teaching post. For a good discussion of the political climate in the universities which relates to Hegel's assessment of Fries see Adriaan Peperzak, *Philosophy and Politics: A Commentary on the Preface to Hegel's Philosophy of Right*, (The Hague: Martinus Nijhof, 1987), pp. 15-31. For the state must take upon itself the task of discerning what is seditious and what destroys the harmony of ethical life. *Philosophy of Right*, p. 9, and para. 270. Hegel claims that the essential principle of the form of the state is thought, and that it was through the state that science was freed from the fetters of religion.

Conclusion

In his attempt to reconstruct the world as an indefinitely extending series of mechanisms, Descartes had grasped the soul as a cognitive foundation only to empty it of all but its metaphysical, cognitive and methodological contributions for grasping the material world. Apart from those conditions the unity of soul and body was essentially a corporeal unity. The question of contiguity of body and soul has always remained an enigma to non-Cartesians, but the Cartesian finds no incoherence as he or she shifts from one optic to another. While the dualism is not contradictory on its own terms it nevertheless created a paradoxical world — a world at once built upon and yet devoid of spirit.

The machine world — governed by a cognitive overseer — seemed to hold out to Descartes the promise of earthly happiness. Such a world was not enough for Kant. Like Rousseau, he believed that such a picture of the world lacked an understanding of human dignity. Building upon the Cartesian innovation of a facultative logic (without duly acknowledging the extent of the debt), Kant sought to save the most important dimension of our being from the encroachment of theoretical knowledge. The terrain of faith became the site of autonomy.

The labour and architectonic of the critical philosophy is a monumental effort to demonstrate the necessary connections between metaphysics, nature and freedom. But, as it has frequently been said, it is based upon a specific set of postulates about nature, and hence a specific theory of nature, which if rendered obsolete, shatters the architectonic and the connections themselves. The task of those who remain impressed by the Kantian turn toward the structure of judgment and who perceive the fruitfulness of the separation between different forms of judgment, whether they be empirical, moral, aesthetic, or metaphysical, has been to find alternative strategies of transcendental legitimation.

Although many of Kant's brilliant contemporaries appreciated the reassertion of autonomy over the spiritless world constructed by the new science, Kant's realm of faith did not reach far enough. For Jacobi and other Romantics all being was to be derived from the realm of faith. Such a derivation entailed the breach with the dualist apparatus that Kant had initially used to locate and legitimate the realm. Fichte and Schelling were both intent on eliminating the dualist residues. But it was Hegel who most consistently followed through the idea of the unity of spirit and who sought to establish that unity not as a postulate of faith (a postulate which could not help but reaffirm the dualist complex it sought to escape from) but as an object of knowledge, as reason itself.

With such a move, the initial Cartesian position is at once overturned, yet reaffirmed and reconstituted. It is overturned in so far as the idea of an ahistorical spirit, even one that is purely of a cognitive nature, is a monstrosity — a *Logic* must have its *Phenomenology*, a form is always a developing content. But it is reaffirmed in so far as the world really is governed by reason, a total system, as Descartes sought to establish. The problem with the Cartesian totality is that it was only conceived from a limited stand-point, the limited stand-point of a science which sought to sever itself from its spiritual moorings. Mechanistic science did have its place in the realm of the spirit, but its place was not that of a master. Hegel's writings on science were too closely allied to the romantic philosophy of nature, too governed by metaphysical concerns that seemed to bear only a nuisance value to scientists working within a quasi-Cartesian environment to survive. It was in the sphere of free spirit, in the social/cultural sphere, that Hegelianism seemed to surpass all previous forms of idealism.

Hegel saw both Descartes's conception of freedom as technical mastery and Kant's conception of freedom as formal autonomy as barbaric in their consequences. The Cartesian technician could, no doubt, find a role in civil society, but what type of citizen would he make, and what would a world be like if all relations were to be subordinated to utilitarian occasions? The Kantian moralist, on the other hand, with his charter of rights and duties was every bit as much a danger to the forms of spirit if he sought to make the world fit into a set of abstract demands. Hegel acknowledged the importance of the rightful demands of the conscience, provided they were subordinate to the social and spiritual patterns that had historically developed and maintained their legitimacy through their provision of substantial personalities.

The distinction between the Kantian and Hegelian conception of freedom remains one of the most crucial points of division in social and political thought. Questions of the relative merits and antagonisms of deontological ethics and communal integration remain at the threshold of political theory.

When we consider the capacity humans have to act upon moral ideals and principles, or reflect on the growing role of the UN and the place in the modern world of human rights' charters and bodies we cannot help but think of Kant. On the other hand, questions of the meaning and end of history which have fed Marxist and more recently American conservative social thought disclose Kantian and Hegelian patterns of thinking. Distinctly Hegelian are the pressing problems of social atomization/fragmentation and the development of forms of substantive social identity, problems which cut far beyond the one dimensional left-right axis that has done so much ideological damage in this century. We have still to come to grips with, to recognize the limits of the scientized world of Cartesianism, and we still are thinking through the problems of autonomy and substantive freedom which are thought, in such different ways, and with such precision by Kant and Hegel.

Index

Adickes, E. 102
Allison, H. 62
Althusser, L. 105
Aristotelian, 9, 12, 21, 26, 27, 33, 42
Aristotle, 9, 12, 42, 106, 121, 122, 165, 171
Augustine, 25, 39
Avineri, S. 187
Baader, F. 161, 182
Bacon, F. 4
Baier, R. 170
Beck, L.J. 38
Beck, L.W. 64, 102, 123
Beiser, F. 146
Berkeley, G. 54
Berkson, W. 42
Berlin, I. 145
Bluhm, W. 35
Bobbio, N. 183
Buchdahl, G. 36, 172
Burke, E. 175, 181
Burman, F. 7, 8, 36
Burrt, E. 33
Butts, R. 98
Caird, E. 97

Capek, M. 171
Carnap, R. 101
Cassirer, E. 49, 59, 62, 64, 69, 97, 106, 107, 108, 120, 147
Caton, H. 35, 37
Christina, Queen. 41
Clarke, S. 75, 100
Cohen, H. 61, 62, 63, 101
Condorçet, A. 32, 41
Cottingham, J. 39
Croce, B. 165
Cullen, B. 184
Curley, E. 38
D'Alembert, J. 31, 41
D'Holbach, P. 42
Descartes, R. vii-viii, 1-49, 54, 57, 59, 60, 106, 109, 120, 129, 167, 191-193
Dickey, L. 184
Diderot, D. 111, 120
Dijksterhuis, E. 33, 42
Dilthey, W. 147, 166, 172, 181

Elizabeth, Princess. 17, 18, 19, 38, 41
Engels, F. 32, 41
Ewald, O. 104
Fabro, C. 40
Fichte, J. 125, 131-142, 145, 147, 149, 150, 155, 156, 161, 168, 169, 173-175, 192
Fischer, K. 61, 62, 96
Fries, J. 161, 169, 189
Galileo, G. 4, 5, 7, 35, 92
Garber, D. 36
Gassendi, P. 13, 14, 15, 42, 43, 129
Goethe, J. 146, 166
Görres, J. 168
Hall, A. 34
Haller, K. 183
Hamann, J. 129, 130, 131, 145
Hardouin, G. 28, 40
Harris, E. 187
Harris, H. 146
Haym, R. 186
Hegel, G. vii-viii, 129, 131, 132, 135, 136, 137, 142-189, 192, 193
Heidegger, M. 62, 101, 143
Heisenberg, W. 107
Helmholtz, H. 107
Henrich, D. 147
Heraclitus, 168
Herbart, J. 187
Herder, J. 129, 130, 131, 145
Hessen, B. 34, 42
Hintikka, J. 36
Hirsch, E. 182
Hobbes, T. 13, 14, 15, 111, 120, 129
Hölderlin, F. 147
Hume, D. 48, 50, 51, 52, 54, 59, 60, 63, 75, 76, 79, 81, 101, 105, 107
Huygens, C. 33, 35, 39
Inwood, M. 172
Jacobi, F. 145, 146, 150, 161, 169, 170, 192

Jolley, N. 59
Kant, I. vii, viii, 36, 37, 47-125, 129-134, 136-43, 145, 146, 149, 151, 152, 155-158, 160, 161, 162, 165-175, 179, 180, 182, 184, 187, 188, 189, 192, 193
Kaufmann, W. 187
Kellenbenz, H. 34
Kepler, J. 4, 92, 171
Körner, S. 97
Kroner, R. 147, 181
Krüger, G. 40
la Mettrie, J. 14, 37
Lachterman, D. 37
Lamb, D. 171
Leibniz, G. 42, 48, 49, 50, 52, 54, 55, 59, 60, 63, 72-76, 95, 97, 99, 100, 101
Locke, J. 48, 49, 50, 51, 52, 54, 59, 60, 74, 100, 105, 184
Loeb, L. 35
Lukács, G. 147, 181, 184, 185
Machiavelli, N. 30, 41, 179
Mainzer, K. 97
Maker, W. 184
Maritain, J. 35
Martin, G. 97, 100
Marx, K. 105, 176, 177, 183, 185
Mates, B. 99
McClaughlin, T. 41
Meinecke, F. 187
Meinecke, W. 99
Mersenne, M. 5, 9, 20, 35, 36, 38
More, T. 121
Müller, A. 168, 175
Nagel, E. 107
Newton, I. 33, 34, 37, 41, 42, 54, 55, 63, 72, 73, 74, 75, 76, 92, 97, 99, 100, 119, 166, 171

Nietzsche, F. 105, 118, 143, 150
Nussbaum, F. 4
Paine, T. 188
Palter, R. 99
Paolucci, H. 171
Parmenides, 168, 169
Pascal, B. 28, 40, 41
Paulus, H. 187
Peperzak, A. 189
Petry, M. 170, 171, 172
Pinkard, T. 170
Plant, R. 184
Plato, 69, 97, 100, 106
Popkin, R. 21, 38
Prauss, G. 62
Regius, H. 13, 14, 15, 16, 23, 39
Reich, K. 104
Rheinhold, K. 146
Riedel, M. 183, 186
Riehl, A. 61, 62, 63, 64
Ritter, J. 186
Robespierre, M. 174
Rosenkranz, K. 147, 186
Rousseau, J.J. 109, 110, 111, 113, 114, 117-123, 167, 168, 173, 174, 175, 191
Röd, W. 35, 36, 38
Saint-Simon, H. 32, 41
Savigny, F. 183
Schiller, F. 121, 131, 143
Schelling, F. 125, 131, 137-151, 155, 157, 161, 164, 166, 169, 171, 172, 192
Schlegel, F. 168, 175, 181, 182
Schleiermacher, F. 161, 175, 181, 182
Schlick, M. 102
Schopenhauer, A. 61, 150
Schouls, P. 41
Schubarth, C. 186
Schultz, J. 69, 97
Sebba, G. 28
Smith, N. 63
Smith, S. 188
Soffer, W. 35
Spinoza, B. 26, 39, 40, 59, 120, 125, 130, 131, 146, 168, 169
Strawson, P. 62
Taylor, C. 145
Vartanian, A. 42
Voltaire, 33
Waszek, N. 184
Weil, E. 186
Weizäcker, C. 64
Wiredu, J. 99
Wolff, C. 101